AF730993

***FM 3-100.21**
(FM 100-21)

Field Manual
No. 3-100.21

Headquarters
Department of the Army
Washington, DC, 3 January 2003

Contractors on the Battlefield

Contents

DISTRIBUTION RESTRICTION: Approved for public release; distribution is unlimited

*This publication supersedes FM 100-21, 26 March 2000

Preface

Contractors have always accompanied our armed forces. However, the increasingly hi-tech nature of our equipment and rapid deployment requirements have significantly increased the need to properly integrate contractor support into all military operations. Recent reductions in military structure, coupled with high mission requirements and the unlikely prospect of full mobilization, mean that to reach a minimum of required levels of support, deployed military forces will often have to be significantly augmented with contractor support. As these trends continue, the future battlefield will require ever increasing numbers of often critically important contractor employees. Accordingly, commanders, staffs, and soldiers must be more familiar with how to plan for and use contractors effectively. This manual, along with an established, formal training program, provides the foundation upon which the Army can promote contractors on the battlefield education.

Field Manual 3-100.21 (100-21) addresses the use of contractors as an added resource for the commander to consider when planning support for an operation. Its purpose is to define the role of contractors, describe their relationship to the combatant commanders and the Army service component commanders, and present their mission of augmenting operations and weapon systems support. This manual is intended for commanders and their staff at all echelons, program executive officers/program managers, and others involved in the planning, management, and use of contractors in an area of operations. It is also a guide for Army contracting professionals and contractors in implementing planning decisions and understanding how contractors will be managed and supported by the military forces they augment.

The prologue introduces contractor support through a notional narrative scenario that demonstrates the process of how contractors would realistically be planned for, supported, and used. Although many contractors support our armed forces, most commanders, staffs and soldiers do not fully understand the differences between a contractor workforce, Department of the Army civilians, and military support units. Accordingly, the manual briefly describes a contract and a contractor, their effect on the Army support structure, and the distinctions between soldier support, Department of the Army civilian support, and contractor support. Additionally, it outlines the governing principles that support the use of contractors and addresses Army policy on contractors on the battlefield in an operational construct

Roles and responsibilities involved in planning contractor support, deploying and redeploying contractor personnel and equipment, and managing, supporting, and protecting contractors are discussed in the manual. It addresses the planning process and relates the planning for contractor support to the military decision-making process. It also addresses the need for a risk assessment that supports the decision process. Additionally, this field manual includes planning considerations unique to contractor support. A discussion on how contractor requirements are integrated into operational plans is also included.

Deployment and redeployment of contractors, including specific guidelines on predeployment processing, time-phased force deployment data development, and reception in the area of operations are presented in detail. Also covered is contractor management during operations, to include a discussion on contractor visibility and accountability and how they will be integrated into the operational support scheme. Included is the support contractors may require from the military in order to adequately perform their support mission, such as facilities, transportation, and life support. Finally, force protection and how contractors will be protected from potential enemy action are also discussed.

This manual reflects relevant doctrine, incorporates lessons learned from recent operations, and conforms to Army doctrine and policy.

The proponent for this publication is the Combined Arms Support Command. Send comments and recommendations directly to Commander, US Army Combined Arms Support Command, ATTN: ATCL-AL 401 1st Street, Suite 227, Fort Lee, VA 23801-1511.

Unless otherwise stated, whenever masculine or feminine gender is used, both men and women are included.

PROLOGUE

Contractors on the Battlefield (a Notional Scenario)

US Central Command alerted the Third US Army (the Army Service Component Command for US Central Command) for another contingency mission as part of a coalition force going into Eastern Africa. Although the operation was primarily peacekeeping in nature, there was a possibility that the situation could develop into open hostilities. This type of operation was not new to the command, they had seen similar action in Somalia and Angola. Because of their past experience, the Third Army's planning staff had developed plans to handle a wide variety of operations, including this newest one in Africa. However, the combatant commander's campaign plan provided for a lean military force to accomplish the mission, requiring his staff to do some creative planning to achieve the right mix of combat and support forces to include the military-to-civilian mix.

The nature of the operation required that the force structure be combat heavy. For the staff, this meant that some of the support force structure would be delayed in deploying, and in some cases left behind. Support requirements, however, still had to include a significant theater force opening package. This package was to execute the aerial port of debarkation and reception, staging, and onward movement support operations as well as to perform the other support functions necessary to support the deployment and begin sustaining the force.

As planning proceeded, it was determined that the US Army Materiel Command's logistics civil-augmentation program umbrella contract was the best option to assist the military early-entry modules and theater force opening package units to operate the aerial port of embarkation and forward support bases, upgrade facilities for long-term occupation, construct base camps, and perform minor construction missions as required. Because of the infrastructure in the region, external support contractors from Egypt could perform several of the supply and transportation functions and provide most of the communications support. These considerations would eliminate the need to deploy a transportation battalion, most of the water purification assets, and most of a signal battalion. Assistant Secretary of the Army for Logistics, Acquisition, and Technology program executive officer/manager system contractors would also be required to provide forward maintenance and technical assistance for many of the Army's latest weapon and battle command systems found throughout the force. Supported units, therefore, needed to ensure that they deployed with sufficient contract administration support to monitor contract performance, which in some cases would be unit personnel.

Although Third Army had used contractors in its past operations, it had never experienced the size and diversity of the contractor presence that was expected for this operation. Planners recognized, because of the scope of the operation and the size of the geographical area involved, that there was a need to have visibility and accountability of contractor employees. Fortunately, Third Army had instituted a program of tracking contractor visibility in accordance with established doctrine. Using many of their primary and special staff members with assistance from the major support units who would be utilizing systems and external contractor support, Third Army was able to identify the contractor employees who would be deploying with them to support this operation. Additionally, direct links to the appropriate contracting organizations for the system contractors through the staff, major support commands, and Army Materiel Command's Logistics Support Element, had been established to ensure theater-specific requirements were added to all contracts.

The Third Army commander directed that all contracts in support of the operation include command directives covering conduct and discipline of contractor employees in the area of operations, and safety and security requirements. The Third Army commander also directed that contractors deploying with the force be integrated into the time-phased force and deployment data as well as receive the requisite predeployment training and processing. This predeployment training and processing would be accomplished at the force-projection platform installations for contractor employees deploying with their habitually supported units and at the designated continental United States replacement centers for those employees deploying as individual replacements. The exception to this policy would be the logistics civil-augmentation program contractor who, because of this program's size and need to be in the area of operations early, would self-deploy and perform its own employee predeployment processing. Although self-deploying, the logistics civil-augmentation program contractor information was still required to be entered into the time-phased force and deployment data, and arrival slots for its aircraft and vessels had to be obtained. Finally, the Third Army commander directed the staff to plan for the operational (facilities, government-furnished equipment and services, and transportation) and life support (including billeting, messing, and medical) required by the contractors supporting the operation.

The essential nature of much of the contractor-provided support and the hostile threat to their operations required that the Third Army commander have his staff conduct a formal assessment of the risks involved in using contractors. The risk assessment included contingency plans to continue essential services provided by contractors through alternate sources (military, Department of the Army civilians, host-nation support, or another contractor) should contractor support become unavailable.

After completing this risk assessment, the staff concluded that the potential conditions of the operational environment posed a medium threat to contractors, but contractor involvement in the operation was critical. To mitigate the risk, the Third Army commander directed that contractors comply with all command safety and security directives and procedures and instructed that the directives be provided to the supporting contracting organizations to ensure that contracts were written to include these or equivalent provisions. Furthermore, military forces would protect contractor operations and movements in those areas where the threat of hostilities was likely, recognizing that the advantageous additional capabilities provided by contracted support may be offset by the diversion of some combat power to protect them. The commander also directed that deployed contractor employees would reside on military compounds. This would assist with security and limit competition for facilities. In accordance with combatant commander policy, the Third Army commander did not authorize the issue of sidearms by contractor employees. Because of these security measures, contractor employees would be permitted to operate in any area that their support was required.

As a result of the decision briefing on the operation, the Third Army commander approved the plans presented by his staff, including a detailed contractor integration plan. This contractor integration plan was put together by the G-3, but was based on information provided mostly by the other primary and special staff members. The Third Army commander then directed that this plan be included as an annex of the OPORD and gave additional guidance to ensure that this annex was read and followed by the various contracting activities supporting the operation.

Finally, the Third Army commander acknowledged the critical need for contractors to support this particular operation and requested that commanders at all levels provide an orientation for their soldiers covering contractors and their relationship to the military forces. "After all," he stated, "contractors have been supporting our forces throughout its history and will continue to do so; we just need to get better at integrating these valued members of our support team into to the overall plan."

Chapter 1

Overview

INTRODUCTION

1-1. Contractors have always supported our armed forces. Numerous examples exist throughout our nation's history, from sutlers supporting George Washington's Army to today's high-tech firms supporting complex weapon and equipment systems. While contractors consistently support deployed armed forces, commanders need to fully understand their role in planning for and managing contractors on the battlefield and to ensure that their staff is trained to recognize, plan for, and implement contractor requirements. Key to understanding basic contracting and contractor management is being familiar with the basic doctrine laid out in this field manual (FM) and FM 4-100.2.

CONTRACTORS AS A FORCE-MULTIPLIER

1-2. Whether it bridges gaps prior to the arrival of military support resources, when host-nation support is not available, or augments existing support capabilities, contractor support is an additional option for supporting operations. When considering contractor support, it should be understood that **it is more than just logistics; it spans the spectrum of combat support (CS) and combat service support (CSS) functions**. Contracted support often includes traditional goods and services support, but may include interpreter, communications, infrastructure, and other non-logistic-related support. It also has applicability to the full range of Army operations, to include offense, defense, stability, and support within all types of military actions from small-scale contingencies to major theater of wars.

1-3. In the initial stages of an operation, supplies and services provided by local contractors improve response time and free strategic airlift and sealift for other priorities. Contractor support drawn from in-theater resources can augment existing support capabilities to provide a new source for critically needed supplies and services, thereby reducing dependence on the continental United States (CONUS) based support system. When military force caps are imposed on an operation, contractor support can give the commander the flexibility of increasing his combat power by substituting combat units for military support units. This force-multiplier effect permits the combatant commander to have sufficient support in the theater, while strengthening the joint force's fighting capability. At the conclusion of operations, contractors can also facilitate early redeployment of military personnel.

UNDERSTANDING CONTRACTOR SUPPORT

1-4. For contractor support to be fully integrated into the operational environment, responsible commanders and their staffs must understand key factors with regard to contractor support.

WHAT IS A CONTRACT?

1-5. Military units receive their guidance and instructions to conduct an operation from published plans and orders, usually operations plans (OPLAN) and operation orders (OPORD). These plans and orders describe the mission and the manner in which the operation will be accomplished. **Contractors receive OPORD-like guidance via their contracts.** A contract is a legally enforceable agreement between two or more parties for the exchange of goods or services; it is the vehicle through which the military details the tasks that it wants a contractor to accomplish and what will be provided to the contractor in return for the goods or services.

WHAT IS A CONTRACTOR?

1-6. Contractors are persons or businesses, to include authorized subcontractors, that provide products or services for monetary compensation. A contractor furnishes supplies, services, or performs work at a certain price or rate based on the terms of a contract. In a military operation, a contractor may be used to provide life support, construction/engineering support, weapon systems support, and other technical services.

Contractor Duties

Duties of contractors are established solely by the terms of their contract—they are not subject to Army regulations or the Uniform Code of Military Justice (UCMJ) (except during a declared war). Authority over contractors is exercised through the contracting officer.

NOTE: For this manual, the term *contractor* refers to the business entity, not the individual employee of the business.

WHAT ARE THE DIFFERENT TYPES OF CONTRACTORS?

1-7. Contractor support is categorized by the type of support that they provide on the battlefield and, more importantly, by what type of contracting organization has contracting authority over them. Battlefield contractors are generally referred to as theater support contractors, external support contractors, or system contractors. Commanders and planners must be aware that a requirement for a particular system or capability may result in the introduction of these type contractors into the operational plan and that contractor management and planning is often significantly different depending on the type of contractor support provided. These differences are described in detail throughout the subsequent chapters of this FM.

THEATER SUPPORT CONTRACTORS

1-8. Theater support contractors support deployed operational forces under prearranged contracts, or contracts awarded from the mission area, by contracting officers serving under the direct contracting authority of the Army principal assistant responsible for contracting (PARC) or other Service/joint/multinational chief of contracting responsible for theater support contracting in a particular geographical region. Theater-support contractors provide goods, services, and minor construction, usually from the local commercial sources, to meet the immediate needs of operational commanders. Theater support contracts are the type of contract typically associated with contingency contracting.

EXTERNAL SUPPORT CONTRACTORS

1-9. External support contractors provide a variety of combat and combat service support to deployed forces. External support contracts are let by contracting officers from support organizations such as United States (US) Army Materiel Command (USAMC) and the US Army Corps of Engineers (USACE). They may be prearranged contracts or contracts awarded during the contingency itself to support the mission and may include a mix of US citizens, third-country nationals (TCN) and local national subcontractor employees. External support contracts include the Logistics Civil Augmentation Program (LOGCAP) administered through USAMC's logistics support elements (LSE), sister Service LOGCAP equivelent programs the Civil Reserve Air Fleet, commercial sealift support administered by the US Transportation Command (USTRANSCOM), and leased real property and real estate procured by the USACE.

SYSTEM CONTRACTORS

1-10. System contractors support many different Army materiel systems under pre-arranged contracts awarded by the Assistant Secretary of the Army for Acquisition, Logistics and Technology (ASA[ALT]) program executive officer (PEO)/program manager (PM) offices and USAMC's Simulations, Training and Instrumentation Command (STICOM). Supported systems include, but are not limited to, newly or partially fielded vehicles, weapon systems, aircraft, command and control (C^2) infrastructure, such as the Army Battle Command Systems (ABCS) and standard Army management information systems (STAMIS), and communications equipment. System contractors, made up mostly of US citizens, provide support in garrison and may deploy with the force to both training and real-world operations. They may provide either temporary support during the initial fielding of a system, called interim contracted support (ICS), or long-term support for selected materiel systems, often referred to as contractor logistic support (CLS).

WHAT IS A REQUIRING UNIT OR ACTIVITY?

1-11. A requiring unit or activity is that organization or agency that identifies a specific CS or CSS requirement through its planning process to support the mission. All requiring units or activities are responsible to provide contracting and contractor oversight in the area of operations (AO), through appointed contracting officer representatives (COR), to include submitting contractor accountability and visibility reports as required.

Requiring units can either be a tactical- or operational-level unit in the AO or a support organization, such as an ASA(ALT) PEO/PM or USAMC, which has identified a support requirement that affects forces in the field. This organization identifies the specific requirements for the support. If it is determined that the requirement is best satisfied by contractor support, this organization prepares the required statement of work (SOW) that supports the contracting process. It should be noted that the requiring unit or activity may not be the organization actually receiving the contractor support. These units are simply referred to as the supported unit.

WHAT IS A SUPPORTED UNIT?

1-12. A supported unit is the organization that is the recipient of support, including contractor-provided support. A supported unit may also be the requiring unit, if it initiates the request for support.

WHAT IS A SUPPORTING ORGANIZATION?

1-13. The supporting organization is that organization or activity tasked to furnish organizational or life support to a contractor. It may be the supported organization, the requiring unit or activity, a functional organization such as a transportation unit or the USACE that would provide equipment or facilities, or any other organization that possesses the support capability. The supporting organization is responsible for providing specified organizational support or life support to contractors as directed. The supporting organization may, if appropriate, use another contractor to provide such support.

WHAT IS A SUPPORTING CONTRACTING ACTIVITY?

1-14. A supporting contracting activity is the contracting organization that supports the requiring unit or activity in arranging contractor support. This organization, staffed with certified contracting personnel, may either be the in-theater contracting organization organic to deploying forces, or contracting organizations from outside the theater such as USAMC, USTRANSCOM, or USACE. Regardless of their location, this organization is responsible for the converting of the requiring unit or activities requirement in contracting terminology, locating potential providers, and negotiating and awarding the contract.

WHAT IS A CONTRACTING OFFICER?

1-15. A contracting officer is the official with the legal authority to enter into, administer, and/or terminate contracts. Within the Army, a contracting officer is appointed in writing through a warrant (SF 1402) by a head of contracting activity (HCA) or a PARC. Only duly warranted contracting officers, appointed in writing, are authorized to obligate funds of the US Government. Active and Reserve Component military personnel, as well as Department of Defense (DOD) civilian personnel, may serve as contracting officers supporting deployed Army forces. There are three types of contracting officers: procuring contracting officer (PCO), administrative contracting officer (ACO), and a terminating contracting officer. The PCO and ACO will be the primary contracting officers commanders will have to work with. Further discussion on contracting officers can be found in FM 4-100.2.

WHAT IS A CONTRACTING OFFICER'S REPRESENTATIVE?

1-16. A COR is an individual appointed in writing by a contracting officer to act as the eyes and ears of the contracting officer. This individual is not normally a member of the Army's contracting organizations, such as the Army Service Component Command (ASCC) PARC office, but most often comes from the requiring unit or activity.

Note: It is important to understand that in some situations, the requiring activity may not be the supported unit. For example, the requiring activity for suport to the FOX chemical, biological, radiological, and nuclear detection vehicle is FOX vehicle PM office. In this case, the PM is responsible to ensure that the supported unit and/or the USAMC LSE provides the requisite CORs to assist in the management of the contract.

1-17. From a doctrinal perspective, each requiring unit or activity should appoint a COR from the supported unit. In all cases, the contracting officer assigns the COR specific responsibilities, with limitations of authority, in writing (see Appendix A for a description of COR guidelines). It should be noted that the COR represents the contracting officer only to the extent delegated in the written appointment and does not have the authority to change the terms and conditions of a contract. Only a PCO or ACO (if appointed) may make changes to an existing contract.

System Contractor COR

For system contractors, the COR would be the individual who ensures that the contractor employee provides the day-to-day management oversight in both garrison and in the AO. This COR is also the individual who ensures that the system-contractor employee is prepared to deploy. In some cases, the system contractor COR would come from the supported unit when the contractor employee has a permanent habitual relationship with the unit. On the other hand, where the contractor employee has an area support mission, the COR would come from the unit, possibly the USAMC LSE, that has area support responsibilities related to the contracted support being provided.

WHAT IS A S

1-18. A SOW defines the government's requirements in clear, concise language identifying specific work to be accomplished and incorporated into the contract. The SOW is the contractor's mission statement. SOWs, prepared by the requiring unit or activity, must be individually tailored to consider the period of performance, deliverable items, if any, and the desired degree of performance flexibility. The work to be performed is described in terms of "what" is the required output rather than either "how" the work is accomplished or the number of hours provided. It also must be understood that any requirements beyond the SOW may expose the government to claims and increased costs.

WHAT IS AN UNAUTHORIZED COMMITMENT?

1-19. An unauthorized commitment (UAC) occurs when there is a purchase agreement that is not binding solely because the government representative (a soldier or Department of the Army civilian [DAC]) who made it lacked the authority to enter into that agreement on behalf of the government. Only a warranted contracting officer is authorized to enter into such agreements. To a businessman especially in a foreign country, anyone that appears to possess the authority to make a purchase on behalf of the US government will sell his or her products or services to that person. Unfortunately, there is no documentation to issue payment. Therefore, the businessman doesn't get paid. This in turn creates resentment towards the US military and can prevent further purchases from being made. Another way a UAC occurs is when a government representative directs the actions of a contractor or their employees through the issuance of a work directive not identified in the SOW resulting in the government incurring additional charges. Both of these UACs can be ratified by a contracting officer; however, the ratification process is lengthy and prevents legitimate purchases from being made in a timely matter.

1-20. Soldiers and DACs can be held financially liable for UACs depending on the circumstances and conduct of the individual initiating the UAC.

CONTRACTOR AND MILITÁRY DISTINCTIONS

1-21. The type and quality of support that a contractor provides is similar to that provided by a military CS or CSS support unit, when considered from a customer perspective. However, there are some fundamental differences that commanders and their staffs must be aware of. These differences include:

- Contractors perform only tasks specified in contracts. "Other duties as assigned" does not apply in a contract environment.
- Contractors and their employees are not combatants, but civilians "authorized" to accompany the force in the field. Authorization to accompany the force is demonstrated by the possession of a DD Form 489 (Geneva Conventions Identity Card for Persons who Accompany the Armed Forces). This status must not be jeopardized by the ways in which they provide contracted support.
- Contractor status as civilians accompanying the force in the field is clearly defined in the Geneva Conventions and other international agreements. They are generally defined as persons who accompany the armed forces without actually being members thereof and are responsible for the welfare of the armed forces. Depending upon their duties, they may be at risk of injury or death incidental to enemy attacks on military objectives. If captured, they are entitled to prisoner

of war status. However, if captured, the contractor's treatment and status will be dependent upon the nature of the hostile force and their recognition, if any, of relevant international agreements.

1-22. Management of contractor activities is accomplished through the responsible contracting organization, not the chain of command. Commanders do not have direct control over contractors or their employees (contractor employees are not the same as government employees); only contractors manage, supervise, and give directions to their employees. Commanders must manage contractors through the contracting officer or ACO. CORs may be appointed by a contracting officer to ensure a contractor performs in accordance with (IAW) the terms and conditions of the contract and the Federal acquisition regulations. The COR serves as a form of liaison between the contractor, the supported unit, and the contracting officer.

CONTRACTOR MANAGEMENT IN THE MILITARY ENVIRONMENT

1-23. The management and control of contractors is significantly different than the C^2 of soldiers and DACs. During military operations, soldiers and DACs are under the direct C^2 of the military chain of command. In an area of responsibility (AOR), the regional combatant commander (previosly referred to as the commander in chief [CINC]) is responsible for accomplishing the mission and ensuring the safety of all deployed military, government civilians, and contractor employees in support of US military operations. The supported combatant commander, through the appropriate ASCC, can direct soldier and DAC task assignment including special recognition or, if merited, disciplinary action. Military commanders do not have, however, the same authority or control over contractors and their employees and only has command authority IAW Department of State rules and regulations. To fully integrate contractor support into the theater operational support structure, proper military oversight of contractors is imperative.

1-24. Currently, there is no specifically identified force structure nor detailed policy on how to establish contractor management oversight within an AOR. Consolidated contractor management is the goal, but reality is that it has been, and continues to be, accomplished through a rather convoluted system that includes the requiring unit or activity, the supported military unit (i.e., normally the Army operational-level support commands, such as logistics, engineer, medical, signal, military intelligence, and personnel), strategic-level commands such as the Defense Logistic Agency (DLA), Defense Contracting Management Agency (DCMA) and/or the USAMC's LSE, that has technical and operational management of the contractors providing specific functional support in the AOR.

1-25. It is important to understand that the terms and conditions of the contract establish the relationship between the military (US Government) and the contractor; this relationship does not extend through the contractor supervisor to his employees. Only the contractor can directly supervise its employees. The military chain of command exercises management control through the contract.

1-26. The military link to the contractor, through the terms and conditions of the contract, is the contracting officer or duly appointed COR, who communicates the commander's specific needs to the contractor, when the requirement has already been placed on the contract. When the requirement is not already placed on the contract, the military link to the contractor is the contracting officer. The contracting officer, not the COR, is the only government official with the authority to direct the contractor or modify the contract. The COR, as stated earlier, who has daily contact with the contractor and is responsible for monitoring contractor performance (see Appendix A) is key to contractor management and control. To manage habitually related system-contractor employees, unit personnel with functional area oversight responsibilities may be trained and designated CORs. This interface allows the contracting officer to monitor and direct the activities of the contractor within the scope of the contract to ensure the commander's requirements are met. For a detailed discussion of contractor management, refer to Chapter 4.

GOVERNING PRINCIPLES OF CONTRACTOR SUPPORT

1-27. The consideration and use of contractors in support of military operations is governed by several principles. The following principles provide a framework for using contractors.

RISK ASSESSMENT

1-28. To properly evaluate the value of contractors to any given military operation, the requiring unit or activity and the supported commander and staff make an assessment of risk. This assessment evaluates the impact of contractor support on mission accomplishment, including the impact on military forces, if they are required to provide force protection, lodging, mess, and other support to contractors, to determine if the value the contractor brings to operations is worth the risk and resources required to ensure its performance. The availability of government support provided to contractors, and any conditions or limitations upon the availability or use of such services, needs to be clearly set forth in the terms of the contract. The assessment also addresses potential degradation of contractor effectiveness during situations of tension or increased hostility. See Chapter 2 for a detailed discussion of risk assessment.

FORCE-STRUCTURE AUGMENTATION

1-29. Contractors supporting military operations can be an additional option for meeting support requirements. The use of contractors enables a commander to redirect his military support units to other missions. Therefore, contractors are force multipliers. Even though contractors are a force multiplier, Army policy is that contractors do not permanently replace force structure and the Army retains the military capabilities necessary to perform critical battlefield support functions. Department of Defense Instruction (DODI) 3020.37 requires the development and implementation of plans and procedures to provide reasonable assurance of the continuation of essential services during crisis situations when using contractor employees or other resources as necessary. If the combatant commander, or subordinate ASCC, has a reasonable doubt as to their continuation by the incumbent contractor, contingency plans for obtaining the essential service from alternate sources, such as military, DACs, host nation, or another contractor, must be prepared and included in the contracting support plan.

1-30. Some units, especially digitized heavy units and the interim brigade combat teams may be supported by a significant number of system contractors necessary to provide technical support to newly released and/or conditionally released equipment. When this is the case, the ASCC must weigh the risk of when and where these system contractors are utilized in the AO.

MISSION, ENEMY, TERRAIN (INCLUDING WEATHER), TROOPS, TIME AVAILABLE, AND CIVIL CONSIDERATIONS

1-31. The consideration of mission, enemy, terrain (including weather), troops (includes contractors IAW FM 3-0), time available and civil considerations (METT-TC) is used to evaluate how contractors are used in support of a military operation. These considerations assist commanders and staff planners in evaluating the risk of using contractors throughout the AO. When the commander determines that the risk associated with using contractors is unacceptable, contractors are not used until the risk is reduced.

INTEGRATED PLANNING

1-32. In order for contractor support to be effective and responsive, its use must be considered and integrated in the deliberate or crisis action planning process. Proper planning identifies the full extent of contractor involvement, how and where contractor support is provided, and any responsibilities the Army may have in supporting the contractor. Planning must also consider the need for contingency arrangements if a contractor fails to or is prevented from performing. Chapter 2 discusses contractor planning in detail.

SUPPORT TO THE CUSTOMER

1-33. Contracts will be written to provide the same quality and extent of support as that provided by military units. The contractor organization must be integrated to minimize disruptions to operations and minimize the adverse impact on the military structure. Links between military and contractor automated systems must not place significant additional burdens or requirements on the supported unit. Although contractors may use whatever internal systems or procedures they choose, adherence to military systems and procedures is mandatory when interfacing with the Army. However, the use of internal contractor systems will not compromise operational information that may be sensitive or classified. Additionally, requiring units or activities must be certain these system interface requirements are reflected in the SOW, contracting officers must do the same in the terms of the contract.

INTERNATIONAL AGREEMENTS

1-34. The international agreements and host-nation laws that apply to the operational area directly affect the use of contractors. They may establish legal obligations independent of contract provisions and may limit the full use of contractor support. Typically, these agreements and laws affect contractor support by—

- Directing the use of host-nation resources prior to contracting with external commercial firms.
- Restricting firms or services to be contracted.
- Establishing legal obligations to the host nation (e.g., customs, taxes, vehicle registration and licensing, communications and facilities support, passports, inter- or intracountry travel, mail, work permits, and hiring of local workers).
- Prohibiting contractor use altogether.

Consideration of these agreements must be made when preparing OPLANs/ OPORDs and contracts. Conversely, the use of contractors must be considered when entering into new or revised agreements.

HABITUAL RELATIONSHIPS

1-35. A habitual relationship is a relationship between a business and the military for habitual support of a unit/organization or system or set of systems. The nature of this relationship is established through the terms and conditions of a contract and extends beyond that of the organization to include the individual contractor employee and soldier. This type relationship establishes a comrade-at-arms kinship, which fosters a cooperative, harmonious work environment and builds confidence in each other's ability to perform.

1-36. The relationship between the Army and some weapon-system contractors may be long-term and continuous. Accordingly, the Army may not be able to deploy these weapon systems without also deploying the supporting contractors. This establishes a habitual relationship that is necessary and appropriate. However, since contractor employees are not actually part of the habitually supported unit, the overall system-contractor management is still the responsibility of the ASA(ALT) and USAMC (for training, simulations and instrumentation only) PM/PEOs through the supported unit CORs as required. See Chapter 4 for a detailed discussion on the management of contractors.

1-37. The responsible functional staff, through the requiring unit or activity or supported unit, ensures that all contractors and their employees are included as part of the unit for deployments and facilitates the accurate and timely incorporation of contractor employees and equipment into the time-phased force and deployment data (TPFDD), even if this effort is in concert with a habitually supported unit. A distinction must be made between contractors who provide direct support to a supported unit and those who provide support on an area basis. When a contractor is in direct support of a specific supported unit, that unit assumes the responsibility for their deployment. For contractors providing support on an area basis, the requiring unit or activity (when not the supported unit) has this responsibility.

1-38. While habitual relationships improve the working relationship between the Army and contractor employees, the contract requirements and the legal relationship between the Army and its contractor are not altered by this habitual working relationship.

POLICY ON USE OF CONTRACTORS

1-39. In the event of emergency or contingency operations, contractors will often be required to perform services in an AO. With this increased criticality of contractor support, especially in the near term within digitized units, the Army's policy (AR 715-9) is that:

- Civilian contractors may be employed to support Army operations and/or weapon systems domestically or overseas. Contractors will generally be assigned duties at echelons above division (EAD); EAD should be thought of organizationally instead of a location on a map. However, if the senior military commander deems it necessary, contractors may be temporarily deployed anywhere as needed, consistent with the terms of the contract and the tactical situation.
- Management and control of contractors is dependent upon the terms and conditions of the contract.
- Contractors are required to perform all tasks identified within the SOW and all provisions defined in the contract. Contractors must be prepared to perform all tasks stipulated in the contract by the government to address potential requirements. Contractors will comply with all applicable US and/or international laws.
- Contractor employees may be subject to court-martial jurisdiction in time of war. (Because federal court decisions limit statutory jurisdiction provisions, commanders should consult their staff judge advocate for specific legal advice.)
- When contractor employees are deployed, the Army will provide or make available force protection and support services commensurate with those provided to DAC personnel and authorized by law.

- Contractors accompanying US Armed Forces may be subject to hostile actions. If captured, a contractor's status will depend upon the type of conflict, applicability of any relevant international agreements, and the nature of the hostile force.

NOTE: A limited number of systems contractors may be required to work on a semipermanent well forward on the battlefield. When this is the case, the ASCC must address this situation during the risk assessment process. In some cases, a formal waiver request to DA may be warranted. See AR 715-9 for furher guidance.

Chapter 2

Planning

"Planning is the means by which the commander envisions a desired outcome, lays out effective ways to achieve it, and communicates to his subordinates his battlefield visualization intent, and decisions, focusing on the results he expects to achieve."

FM 3-0

INTRODUCTION

2-1. The Army's primary mission is to deter war and, if deterrence fails, to fight and win. Contractors play a vital role in the Army's ability to accomplish and support this mission. Also, contractors provide a responsive alternative to increasing the number of support forces necessary to perform the mission. Planning for contractor support is part of the overall support plan for any operation. To be effective, contractor support must be considered early in the planning process and continuous throughout the operation. Planning addresses how and where the support should be provided. The planning process also includes evaluation of the risks involved and determination of the extent that contractors should be supported by the military. Planning establishes the basis of requirements for support by a contractor. If contractor-support requirements are not addressed during planning, the SOW and support requirements that must be communicated to a contractor through a contract are either omitted or included too late, thereby reducing the effectiveness of a contractor's ability to effectively support the mission.

ROLES AND RESPONSIBILITIES

2-2. Commanders and staffs at all echelons have the responsibility to ensure all support, to include that provided by a contractor, is adequately considered and integrated in the planning process. Because the planning process is a coordinated staff procedure used by a commander to determine the best method of accomplishing assigned tasks and to direct the action necessary to accomplish his mission, every echelon has certain planning responsibilities. The geographical combatant/subordinate unified commander and his staff initiate the process and determine the requirements. Subordinate levels then refine the plan to meet their specific needs. To ensure contractor support is fully planned for, several commands, staffs, and activities, including the supported combatant commander, joint task force (JTF) staff, supported ASCC, Army component to the JTF Army force (ARFOR), supporting ASCCs, requiring activity, functional staffs/organizations, as well as the Army contracting activities, must be included among the key planners. Additional contracting agencies that are often involved in contracting and contractor-related planning are DLA and DCMA.

JOINT FORCE COMMANDERS

2-3.The geographical combatant (hereafter referred to as combatant commander unless a specific functional combatant commander is identified)/subordinate unified command commander (if applicable) and, in most cases, a JTF commander and their staffs prepare plans to conduct joint operations as directed by the Joint Chiefs of Staff. The combatant commander is authorized to plan for, deploy, direct, control, and coordinate the actions of assigned and attached forces. Joint operations planning at this level requires significant coordination and effort. Joint planners make extensive use of the staff study, estimate of the situation, operations analysis, and intelligence preparation of the battlefield (IPB) and logistics assessment (referred to as logistics preparation of the theater in Army doctrine) in order to provide the commander the information necessary to determine alternative courses of action and to make his decision on how the operation will be supported. It is the combatant command that establishes the theater policies and guidance for the use of contractor support. This guidance will include, but is not limited to, restrictions on contractor support (by area, phase of operation, or other measures as appropriate), contractor-deployment and theater-entrance policies, and general order applicability to include force protection and security policy for contractor employees.

SUPPORTED ARMY SERVICE COMPONENT COMMANDER AND ARMY FORCE COMMANDER

2-4.The supported ASCC commander prepares, trains, equips, administers, and provides Army forces to the JTF. The ARFOR, which could be the ASCC itself in some operations, conducts Army operations to support the combatant commander and JTF objectives. CS and CSS are normally Service responsibilities unless otherwise directed by the combatant commander. Following the planning guidance provided by the combatant commander regarding contractor support, the ASCC commander is responsible for determining how this guidance is executed by the ARFOR. It is at this level that the specific Army contractor management and support responsibilities are assigned. The ASCC staff, following the G-3's lead, is responsible for integrating contractor operations into their portions of the OPLAN/OPORD.

SUPPORTING ASCCS AND MACOMS

2-5.The supporting ASCC, typically US Army Forces Command (USAFORSCOM), US Army Europe (USAREUR), or US Army Pacific (USARPAC), along with USAMC, is responsible for providing the requisite resources to the supported ASCC necessary to conduct full-spectrum operations. In this capacity, these commands have a planning responsibility to ensure that the myriad of details related to deploying and supporting the providing forces are addressed. In many cases, the forces being provided bring with them a supporting system or possibly external support contractors. When this occurs, the supporting ASCCs and/or major Army command (MACOM] must articulate a supporting contractor employee presence, so that the contractor can be integrated into the deployment/redeployment, accountability, visibility, and support structures. These supporting commands must closely coordinate with the supported ASCC to ensure that all contractor resources are provided in a timely manner and IAW the combatant commander, JTF, and supported ASCC/ARFOR plans and policies.

REQUIRING UNIT OR ACTIVITY

2-6. As discussed in Chapter 1, the requiring unit or activity is that organization or agency that identifies a specific CS or CSS requirement through its planning process to support the mission. It may be a tactical- or operational-level unit having a specific support requirement for an operation or a PEO/PM responsible for a major system. When it is determined that contractor services will be utilized to fulfill an activity's requirement, the requiring unit or activity, through its COR, is responsible for integrating their portion of the contractor support into the ASCC/ARFOR operational plan. This includes such things as identifying and planning deployment/redeployment requirements, force protection needs, as well as government-furnished equipment (GFE) and life support. The requiring unit or activity must consider the cost (personnel, equipment, materiel, and funds) involved in providing the necessary support and identify any beyond its capability. The requiring unit or activity is the organization responsible for preparing the SOW when contractor support is used. Additional requiring unit or activity planning responsibilities are discussed later in this chapter.

CONTRACTING STRUCTURE

2-7. FM 4-100.2 addresses the Army contracting structure at the various operational and tactical levels and describes its role in contractor-provided support. The members of this Army contracting structure, including the combatant commander's logistics procurement support board (CLPSB), are key participants in operations planning and provide technical advice and guidance during the requirements-determination process for the use of theater support and, sometimes, external support contractors. It is essential that the members of the contracting structure be included in all planning to ensure the aspects of contractor support are identified and considered. Once contractor support is identified as the preferred source of support, the PARC and other contracting oversight organizations such as DCMA, DLA, USACE, and USAMC subordinate commodity commands, through the supporting contracting activity, are responsible for translating specific planning guidance into contract language. When system contractors are involved in supporting an operation, the contracting activity that supports a particular PEO/PM-managed system must be involved in the planning to ensure that necessary operation-related provisions are incorporated into the governing contract. The USAMC LSEs often serve as the liaison between the supported unit and the contracting activity to communicate operational requirements.

PLANNING PROCESS

2-8. Planning for contractor support is an integral part of the planning for any operation and follows the same process as any other aspect of a military operation. Planning for an operation, whether deliberate or crisis action, requires the integration of combat, CS, and CSS capabilities. For contractor support to be successful, it must be formally integrated early in the deliberate planning process to ensure that it is adequately considered.

Significant contractor planning in a crisis-action situation is problematic at best due to the short timeframe available. Regardless, the planning must identify and include specific requirements and responsibilities in the operational plan.

MILITARY DECISION MAKING

2-9. The foundation of planning is the military decision-making process (MDMP). MDMP is a single established and proven analytical seven-step process (see Figure 2-1). It is an adaptation of the Army's analytical approach to problem solving. The MDMP is a tool that assists the Army commander and staff in developing estimates and a plan, by helping them examine the battlefield situation thoroughly, knowledgeably, and with clarity and sound judgment. The full MDMP is a deliberate, sequential, and time-consuming process used when adequate planning time and sufficient staff support are available to thoroughly examine friendly and enemy courses of action. However, the MDMP process also applies in a time-constrained environment. An abbreviated discussion of the MDMP as it relates to planning for contractor support is presented below. For a full discussion of MDMP, refer to FM 5-0.

Mission Analysis

2-10. When a new mission is received (first step in MDMP), the commander and staff assemble the necessary planning tools and analyze the mission (second step in MDMP). Mission analysis is critical to the MDMP. It initiates the commander's battlefield visualization. Mission analysis defines the tactical problem and determines feasible solutions. MDMP mission analysis consists of 17 substeps, not necessarily sequential. This manual discusses only those substeps that have a direct relationship to contractor-support planning.

2-11. A review of available assets early in the mission-analysis process adds and deletes from the current task organization and examines support relationships and the status of all units. The commander and his staff consider the relationship between specified and implied tasks and available assets. From this they determine if sufficient assets are available to perform all the specified and implied tasks. If there are shortages, they identify additional resources needed for mission success, including possible contractor support.

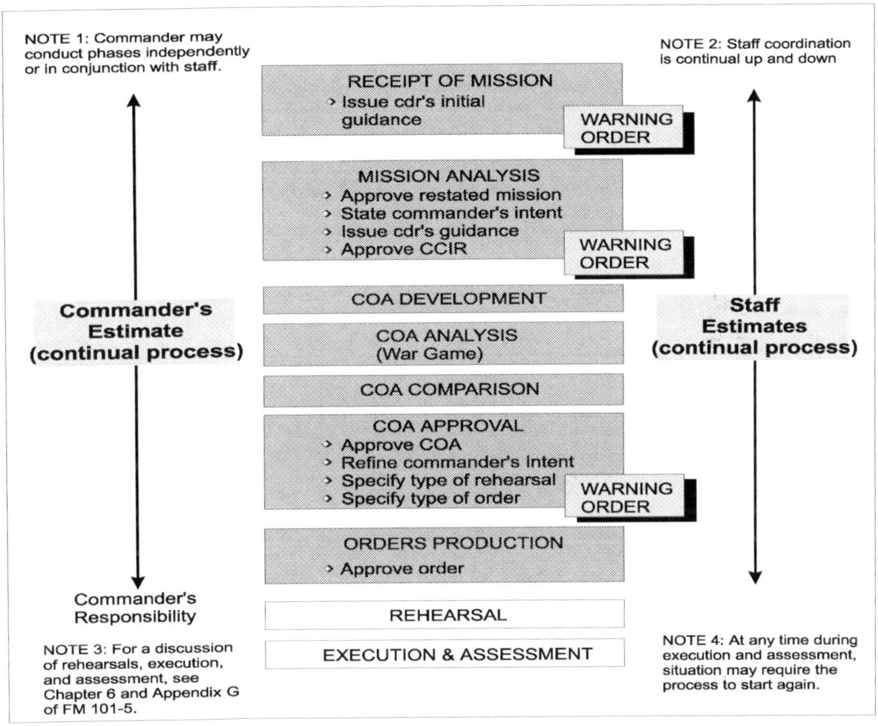

Figure 2-1. The Military Decision-Making Process

2-12. In addition to a review of available assets, two other mission-analysis steps pertain to contractor support. The first is to determine constraints. Constraints can take the form of a requirement to do something (for example, maintain a capability to provide life support to all Services at the aerial port of debarkation [APOD]) or to prohibit an action (for example, a diplomatically imposed limit on military personnel in the area of operations). Planners must identify and understand these constraints which, combined with the review of available assets, may increase the need for additional resources or require the use on nonmilitary support such as contractors.

2-13. The other step under mission analysis pertaining to contractor support is to conduct a risk assessment to identify the tactical risk hazards presented by the mission. This has a direct bearing on contractor support as it relates to whether or not to use contractors or, if contractors are used, to what extent force-protection measures are needed to protect contractor operations and personnel from the hazards. Commanders at all levels need to pay particular attention to those limited number of system-contractor employees whose specialty does not have a military equivalent.

COURSE OF ACTION DEVELOPMENT

2-14. Following the analysis of the mission, planners develop courses of action for analysis and comparison. Each course of action developed must meet the criteria of suitability, feasibility, acceptability, distinguishability, and completeness. A good course of action positions the force for future operations and provides flexibility to meet unforeseen events during execution. The initial array of forces is developed during this step. The individual designation of units is not done at this point, rather units are grouped by type. During this step the potential use of contractor support should be considered and in some cases identified. As stated earlier, some weapons and ABCS and STAMIS systems may require the support of a significant number of system-contractor-support personnel. Planners must be fully aware of this possibility because of the impact it will have on other planning aspects. Additionally, when the number of forces required is greater than the total military support available, the use of contractor support to fill that shortfall must be explored.

COURSE OF ACTION ANALYSIS AND COMPARISON

2-15. Following the development of courses of action, an analysis and comparison of those courses of action will determine which one accomplishes the mission with the least risk. During these steps of MDMP, planners must pay particular attention to support relationships and constraints and to the impact contractor support would have on mission accomplishment. It is during this step that issues related to contractor support are explored, including cost, deployment, operational and life support, force protection, and location on the battlefield. Planners must become fully aware of the risk and impact contractor support will have on the operation. Critical analysis, which includes potential contractor support, will give the commander the necessary information upon which to make a decision regarding the extent contractor support will be utilized.

COURSE OF ACTION APPROVAL

2-16. Knowing the results of the course of action analysis, the commander decides on the course of action he believes to be the most advantageous. He then issues any additional guidance on priorities for CS and CSS activities (particularly for resources he needs to preserve his freedom of action and to ensure continuous support), orders preparation, rehearsal, and preparation for mission execution. This step in the MDMP is where specific decisions regarding the use of contractor support are made and then incorporated into the operational plan. Once the decision to use contractors in support of the operation is made, the specific details regarding contractor support (funding, deployment/redeployment, operational and life support, force protection, and location on the battlefield) must be incorporated into the appropriate OPLANs and OPORDs. Failure to include this important information hinders the communication of the decisions to those who must implement and execute the contractor-support scheme.

RISK ASSESSMENT

2-17. Contractor support of military operations involves a degree of risk. Contractors have traditionally supported the Army during numerous operations, to include combat operations, and can be expected to do so in the future. The decision to use contractors in an area of operations requires an assessment of the risks posed to the contractor and his employees and the potential impacts on the operation itself. Commanders must consider the difficulties facing contractors when hostile action against them is likely. If failure of the contractor to provide the required support could jeopardize the overall success of the operation, contractor support may not be suitable. The primary areas of concern include responsiveness of support, transitioning from peace to war, continuation of essential services, and organic capability, if it exists. Likewise, commanders must consider the risk that a contractor posses to the operation, in terms of the potential for sabotage, or other intentional overt or covert action from the contractor's employees.

2-18. A risk assessment considers essentiality or the impact on the military mission; alternatives to contractors, which look at active/reserve military forces and DAC capabilities, other Services, multinational forces, and host-nation support; and danger to the individual. For example, system contracts for technical assistance and sustainment are critical to readiness, with few or no alternatives to contractors. Conversely, service contracts (e.g., messing, laundry, or sanitation) affect living standards, but not mission accomplishment (at least in the short run). Finally, commanders must understand that contractor personnel are not soldiers; they might refuse to deliver goods or services to potentially dangerous areas or might refuse to enter hostile areas regardless of mission criticality.

2-19. In addition to operational risks, there may be risks unique to using contractors. Their work must be funded. The contractor management structure is often not as clear as C^2 of military units. Contractors may not be able to secure subcontractors, facilities, material, real estate, transportation, or utilities or, if they are able to, only at a prohibitive cost. There may also be US, host-nation, and even third country legal or regulatory constraints on the contractor support. For example, contractor-acquired supplies brought into a theater may be subject to potentially prohibitive excise taxes even though military supplies are not. The key to avoiding this situation is the early involvement of support judge advocate general (JAG) staff in the contractor planning process.

2-20. Units/organizations or activities requiring support on the battlefield must carefully draft the requirements to specify the services and conditions under which they are required. Potential contractors must be fully aware of what is involved. When contractors are willing to perform under dangerous conditions, the cost of a contract may be substantially influenced by the risk the contractor is being asked to accept. Additionally, contractors may be willing to perform under dangerous conditions if the Army meets specified security requirements by providing escorts, training, protective clothing and equipment, or site security to ensure their protection and safety. Commanders accept responsibility for the security of contractor personnel when contractor support is used. Chapter 6 provides more details on contractor force protection.

2-21. The other major area of risk assessment is the real possibility of direct or indirect actions taken against US forces by contractor employees or individuals posing as contractor employees. While not a direct focus of this manual, the risk of utilizing theater support contractors in lieu of US or coalition military support is an important consideration for all commanders and their staff planners. See FM 3-100.14 (100-14) and FM 101-5 for more discussion on risk management and performing risk assessment.

CONTRACTOR PLANNING CONSIDERATIONS

2-22. Contractor support should not be considered just another support option when planning military operations. There are some very unique planning considerations involved when planning for this type of support. Proper planning should, however, make contractor-provided support reasonably transparent to the end user. Planning for contractor support identifies the full extent of contractor involvement, how and where contractor support should be provided, and any responsibilities the Army may have in supporting the contractor. Also, the need for contingency arrangements if a contractor fails to or is prevented from performing must be considered. Special consideration must be given to system-support requirements where no military support is currently available.

Contracting Authority

2-23. There can be numerous sources of contractor support for an operation, each supported by a variety of separate contracting activities, each with its own contracting authority. These may include the PARC/theater contracting command, support-unit contracting staff, USAMC, USACE, DLA, DCMA, USTRANSCOM, the Intelligence and Security Command, and others. When planning for theater, and sometimes external, contractor support, planners must be aware of the operational principle of centralized contracting management to achieve unity of effort and to prevent individual elements from competing for the same resources. FM 4-100.2, Joint Publication (JP) 4-07, and JP 4-08 address this operational principle in further detail. With several contracting activities influencing contractor support in an AO, planners must recognize—

- First, that they exist
- Second, that they must be linked to achieve central management.

2-24. This is achieved through the conscious designation, in OPLANs and OPORDs, of an organization overall responsible for contracting in a theater, usually the CLPSB, Joint Acquisition Review Board (JARB), or the Army's Acquisition Review Board (ARB). By doing so, the commander's decisions and guidance regarding contractor support can be communicated to all involved contracting activities so that a harmony of effort may be achieved. These boards—

- Prioritize requirements against available funding.
- Consolidate requirements from requiring activities to form one contract instead of several.

- Allocate support so that the military and contractors are not competing for scarce resources such as local transportation, real estate, facilities, and utilities.
- Determine the source of contractor support: whether to use a theater support or external support contractor.

Contractor Employee Accountability

2-25. Contractor accountability has been, and continues to be, a significant challenge to commanders at all levels. While contractor employee accountability is a personnel (G-1) function, operational specific contractor accountability responsibilities and procedures must be identified and codified early in the planning process. Because of its inherent responsibilities to its citizens, the US government provides certain support that it is not obligated to do for other nationalities. When US citizen contractor employees are involved in supporting an operation, they must be accounted for in the similar manner as military and DAC personnel. Accordingly, specific US citizen and selected TCN contractor employee data will be required and reported through the tactical personnel system (TPS) by the requiring unit or activity. For additional information on contractor employee accountability, refer to Chapter 4.

Contractor Visibility

2-26. The overall contractor presence, along with its activities and movement, must be orchestrated with those of military forces. Contractor visibility couples contractor employee accountabiltiy information with additional contract-related information to include how many contractors companies there are supporting the opeation, where they operate, when they provide support, the criticality of their support as well as their command-support mission and relationships. Contractor visibility information is necessary for the ASCC G-2, G-3, G-4, provost marshal, surgeon, and PARC to complete their specific planning requirements. Additionally, the ASCC, via the ARFOR, may provide some military support to Army system- and external support contractors as well as some theater support contractors, to include operational and life support and force protection, which dictates a need to quantify the total number of contractors supporting military operations.

2-27. Currently, there are no standard procedures nor automated systems specifically designed to capture visibility of the myriad contractors on the battlefield. Furthermore, there is no one lead staff element responsible for establishing and maintaining contractor visibility. Therefore, it is imperative that mission specific responsibilities and procedures be identified and codified early on in the planning process. For a detailed decussion on contract visibility, see Chapter 4.

Deployment/Redeployment

2-28. Contractor deployment and redeployment is a major consideration for all military operations, but it is highly dependent on METT-TC, as well as the type of contractor providing support. The G-3 staff must be fully cognizant of contractor deployment and redeployment requirements. By definition, theater support contractors do not deploy or redeploy to and from the AO. System and external support contractors on the other hand, will have to deploy some or all of their equipment and personnel into the AO, except when they are already forward deployed, such as in Korea and South West Asia, or when they hire local nationals as part of their contract.

2-29. A very important planning consideration is how these syste- and external support contractors actually deploy and redeploy. In a non-restricted theater, they may self-deploy utilizing commercial assets. In a restricted theater they most likely will deploy utilizing military-deployment processes and strategic-lift assets. The how and when contractors enter the theater must be planned and stipulated in OPLANs and OPORDs to ensure supporting contracting activities can incorporate deployment and redeployment requirements into contracts and to allow the contractor to efficiently plan for and prepare for movement of his personnel and equipment. Failure to identify and coordinate the method of arriving in the AO may interfere with the organized flow of forces into a theater and contractor support that is not in place when required. See Chapter 3 for a detailed discussion on contractor deployment and redeployment.

Location on the Battlefield

2-30. In today's operational environment, contractors can expect to perform anywhere in the AO, subject to the combatant commander's risk assessment and the terms and conditions of the contract. Both the G-2 and G-3 staffs need to be aware of contractor location and criticality advise the commander on issues on when and where contractors will provide support on the battlefield.

2-31. As stated in AR 715-9, civilian contractors may be employed anywhere in the AO necessary to support operations and weapon systems. Generally, contractors work at the EAD-level; however, if the ARFOR commander determines that their services are required at lower echelons, they may be temporarily deployed as far forward as needed, consistent with combatant commander's policy, the tactical situation, and the terms and conditions of the contract. Refer to Chapter 4 for additional information on contractor location within the AO.

Force Protection

2-32. Force protection measures for contractor support must be based on battlefield-location decisions made by the combatant commander and subordinate joint and Army commanders. Protecting contractors who are in direct support of Army forces on the battlefield is the ARFOR commander's responsibility via the G-3 staff. When contractors perform direct support of Army forces in potentially hostile areas, the supported military forces must assure the protection of the contractor's operations and personnel. Commanders and planners must determine the need for contractor force protection early in the planning process and identify forces to provide security. Mission, threat, and location determine the degree of force protection needed. To maintain force protection for contractors, the ARFOR commander, the supporting contracting officer, and the responsible requiring unit or activity must establish and utilize procedures to identify contractors and their status and location within the AO. Chapter 6 provides a detailed discussion on contractor force protection to include provisions that require contract employees to comply with a combatant commander's force-protection rules.

Status of Contractor Employees

2-33. Contractor employees are neither combatants nor noncombatants. Under international agreement, they are considered civilians authorized to accompany the force in the field and, as such, they should not be consciously placed in a position where they might be subject to intentional attack. Certain actions on the part of a commander may unintentionally compromise this status and place them in the combatant category. Generally, the three conditions that make an individual a combatant are—

- Being commanded or controlled by a published chain of command.
- Wearing a distinctive insignia or uniform.
- Openly carrying of arms.

2-34. Although the first condition is not easily determined visually by an enemy, if the commander permits contractor employees to wear military-looking uniforms and carry weapons, he may jeopardize their status as civilians authorized to accompany the force in the field. Commanders, therefore, must carefully consider their decisions regarding the use or location of contractors in the theater of operations. In some cases, a source of support other than contractors may be more appropriate.

Extent of Government Furnished Support

2-35. Contractors are expected to be self-sufficient, handling all actions necessary to perform under the conditions of the contract without additional assistance from the government. However, in some circumstances, it is appropriate and necessary for the government to provide support. The extent of government-furnished support to a contractor must be identified and coordinated with the appropriate G-4 during the planning process. Early identification of requirements allows the G-4 to coordinate the needed support and to allow contracting activities to communicate the requirements to the contractor through the contract. Chapter 5 addresses this subject in detail.

Funding

2-36. Although cost should not be the overriding factor in determining the use of contractor-provided support, funding must be identified and earmarked during the planning process. Federal law requires that all contracts be funded, which means that at the time of contract award, funds are available to cover the total estimated cost of the contract. Exceptions to this law are identified in the Defense Federal Acquisition Regulation Supplement. Therefore, the resource management staff must be involved in the planning process and must provide accurate information regarding the availability of funds. See FM 4-100.2 for more information regarding contract funding.

OPERATIONAL PLANS

2-37. Operational plans are a means by which the commander expresses to his subordinates his battlefield visualization, intent, and decisions, focusing on the results the commander expects to achieve and his vision of the end state of an operation. These plans bridge the gap between the present and the future by answering in advance who will accomplish what future actions and when, where, why, and how they will be accomplished. Operational plans are the critical link between the commander and the organizations that must implement the commander's decisions. They also communicate the commander's decisions to the contracting activities so they can be incorporated into contracts.

CONTRACTING SUPPORT PLANNING

2-38. Although not directly addressed in FM 5-0, it is common practice to have a contracting support plan in most current OPLANs/OPORDs. The contracting support plan lays out the operation-specific contracting procedures, responsibilities, and actions. It begins with the combatant commander's guidance, is written by the theater PARC in close coordination with the G-4 and other staff planners, and is normally found as an appendix to the logistics annex. It serves as the mechanism for providing detailed guidance on contracting support for a specific military operation and covers the function of acquiring (contracting for) theater support contracting for a particular operation. It normally does not include discussion on external- and system-support contractors.

2-39. The contracting support plan ensures that theater-contracting personnel plan, prepare, and coordinate to support deployed forces. It ensures that theater support contracting plans and procedures are known and included in appropriate portions of the OPLAN/OPORD. Contracting professionals use this plan to properly integrate the function of contracting into the concept of support, ensuring host-national support, contingency contracting, and other support options are properly included and time-sequenced in all support planning.

CONTRACTOR INTEGRATION PLANNING

2-40. Numerous lessons learned related to contractor support to military operations clearly identifies the need to better integrate contractors into the military-planning process. Detailed contractor integration planning **(not to be confused with the contracting support plan discussed above)** is necessary to addresses specific contractor-related deployment, management, force protection, and support requirements that are routinely identified, but not well articulated, in recent operational planning. One way to address this need is to develop and publish a contractor integration plan as a separate annex to the OPLAN/OPORD. While not specifically addressed in current doctrine, FM 5-0 allows for the commander to determine the type and number of annexes in each OPLAN/OPORD. **In any case, specific contractor integration information must be addressed in the applicable annex and/or appendix of the OPPLAN/OPORD.**

2-41. The ASCC commander's operations officer, with assistance from the primary and special staffs, is responsible to ensure that the contractors are properly integrated into to the overall plan. At a minimum, the individual portions of the OPPLAN/OPORD must address how contractors supporting an operation are be managed, deployed, supported, and protected. More specifically, the OPLAN/OPORD, with or without a separate contractor integration plan annex, must provide the following:

- G-1 input on contractor personnel support (mail, legal, etc.), contractor employee theater-entrance requirements as well as accountability reporting requirements.
- G-2 information on contractor employee clearance and security procedures.
- G-3 information on contractor deployment/redeployment.
- G-4 guidance on the issuance of GFE and life support.
- Provost Marshal developed force protection policies and procedures.
- SJA information on legal issues.
- Surgeon-stipulated special contractor medical requirements.
- Staff engineer guidance on facility use.

2-42. The operations office must ensure that the staff planners conduct advance planning, preparation, and coordination to incorporate contractor support into the overall operation. The OPLAN/OPORD should encompass all types of contractor support (theater support, external support, and system), originating from anywhere in the world and serve as the critical link between the supported ASCC, the various functional support elements, contracting activities, and the supporting contractors. **With or without a separate contractor integration plan annex, the OPLAN/OPORD must clearly communicate operational specific contractor integration requirements to the contracting activities so that tailored contracts may be executed.** Appendix B contains a sample contractor integration plan annex.

SPECIAL CONSIDERATIONS

2-43. Contractor-provided support brings with it special considerations that do not normally pertain to other types of support. These considerations affect how contractors provide support and prepare for and deploy to an AO. Planners must ensure that they are addressed and included, when appropriate, in OPLANs/OPORDs.

RELATIONSHIPS

2-44. The type of working relationship between a contractor and the military determines any specific military training for the contractor and the processes for deployment, redeployment, and support in an AO. The relationship is either habitual or nonhabitual and is established through the terms and conditions of the contract.

Habitual

2-45. A habitual relationship is a long-term relationship, normally between a system contractor who has a dedicated or direct-support (vice area-support) relationship to a specific supported unit. This type relationship may extend beyond the organization to include the individual contractor employee and soldier. It establishes a comrade-at-arms kinship, which fosters a cooperative, harmonious work environment and builds confidence in each other's ability to perform. The existence of a habitual relationship greatly facilitates the planning for predeployment processing, deployment/redeployment, operational and life support, and force protection by incorporating the needs of the contractor with those of the unit being supported.

Nonhabitual

2-46. A nonhabitual relationship manifests no established, long-term business relationship between a contractor and a specific supported unit or organization. This occurs when support is general (area), rather than direct support. This could also occur when the required support is short term or immediate in nature where there is no time to establish a habitual relationship. Nonhabitually related contractors are clearly more difficult to plan for; special staff and command actions may be necessary to ensure proper planning for specific contractor-related deployment, management, force protection, and support requirements. Special actions may include identifying a support unit to take responsibility to coordinate the plan to deploy and manage these general-(area) support contractors.

TRAINING WITH CONTRACTORS

2-47. The concept of contractor support must be embedded in the Army's training hierarchy. Military, civilian, and contract personnel must be trained to effectively perform their duties. Commanders and staff planners must be familiar with the government responsibilities for supporting contractors: management and accountability, deployment/redeployment, operational and life support, and force protection. Training with contractors must be written into appropriate mission training plans and training support packages, and included in situational training exercises and field training exercises.

2-48. Contractors should be included and funded to participate in selected training events as well as exercises and other collective training events. By doing so, they can develop their mobilization and general military skills that may be required for a deployed environment, including nuclear, biological, and chemical (NBC) training; use of communications equipment; individual field sanitation skills; survival, escape, resistance and evasion (SERE); military unit organization; and, if applicable, weapons familiarization and other common task training. Rotation with maneuver forces to training sites for exercises mutually benefits contractors and units and should be part of the ongoing peacetime activity for contractor personnel. Contractor support should be written into training scenarios.

RESPONSIVENESS OF SUPPORT

2-49. The operational environment and METT-TC determine how responsive the contractor can be in fulfilling the terms of the contract. When assessing risk, commanders must evaluate factors affecting contract performance that are not under the control of the contractor, such as transportation assets needed to move supplies or force protection.

2-50. The establishment of a flexible, functioning, deployable network of properly trained CORs greatly facilitates responsiveness of contracting officers and a contractor's ability to meet contract requirements. When this structure is not in place, performance is greatly reduced, wasting valuable resources as contracting officers wait to approve and process changes and new requirements. The structure used to manage contractors on the battlefield should be trained and exercised. The resulting experience will enhance the planning process significantly.

CONTINUATION OF ESSENTIAL SERVICES

2-51. While more directly focused on contracting vice contractor integration planning, the continuation of contractor essential services must be addressed during risk assessment. To ensure continuation of essential services, DOD (in DODI 3020.37) has instructed that:

- Contractors who provide essential services should continue those services, under the terms and conditions of the contract, during periods of crisis until released or evacuated by military authority.
- The Army will develop and implement plans and procedures that reasonably assure continued service during crisis situations, using contractor employees or other resources as necessary.
- The ASCC commander will prepare a contingency plan for obtaining the services from alternate sources for situations where there is reasonable doubt that essential services provided by a contractor will not continue.

2-52. Particularly when operations may transition to a hostile environment, advance planning is essential to identify a backup source of support and the resources necessary to enable the contractor to continue or accept the risk if the support is not provided.

COMMUNICATING CONTRACTOR REQUIREMENTS

2-53. The OPLANs/OPORDs, with or without a separate contractor integration plan annex, must describe the scope of contractor support as well as contractor integration requirements. This information then becomes the basis for developing contract requirements. The contract and its supporting documentation define all requirements for the contractor. Likewise, OPLANs/OPORDs must provide the same information to the units receiving the support. Commanders, staffs, and contractors must understand that the contractor is not legally obligated to meet any requirement (deployment, force protection, life support, or in-theater management) not contained in the contract; without a requirement specified in the contract, the government has no basis for directing or requiring any contractor action. If the government directs the contractor to meet requirements that are beyond the specifications of the contract without proper modification, the government can expect to be billed/charged, or the contractor may refuse to meet the requirements. If funds are not available, a violation of the law may occur.

2-54. All requirements for contractor support originate in a government SOW that describes the parameters (what, where, and when) of the requirement, government support to be provided (such as transportation, security, and life support), and the restrictions and control measures that apply to the contractor. The SOW, along with terms and conditions, becomes the contract for the support requested.

OPERATIONS PLAN/CONTRACT INTERFACE

2-55. Written properly, OPLANs/OPORDs serve to communicate contractor-support decisions and responsibilities throughout the force. As the focal point for these decisions, the OPLAN/OPORD and its appropriate annex informs the supporting contracting activities of the requirements to include in contracts. Contracting professionals translate the commander's decisions into contract language, making them legally binding for the contractor performing the work. Once again, if a requirement is not included in a contract, the contractor is not obligated to comply. Ideally, the planning for contractor support should be accomplished, to the extent possible, during the deliberate planning process so there is adequate response time for contracting professionals to translate requirements into contracts. When sufficient time is not available, planners must still ensure that they communicate, as quickly as possible, contractor-related requirements to the supporting contracting organization.

OPERATIONAL SPECIFIC CONTRACT LANGUAGE

2-56. The requiring unit or activity identifies and initiates the requirement for contractor support. From details laid out in the appropriate OPORD/OPLAN, the requiring unit or activity may have to develop new requirements to provide to contracting professionals in the supporting contracting activity. The supporting contracting activity is responsible to translate the commander's decisions into contract language, making them legally binding for the contractor performing the work. For example, the requiring activity for a specific system contract may be an ASA[ALT] PEO/PM office. This PEO/PM office would typically communicate specific operational requirements to its supporting contracting activity in the form of a SOW and identify what is needed, when, and where. Included in this identified requirement is information related to any government-furnished support, materials or equipment that may be provided, as well as the standards for measuring the expected quality and acceptability of performance. This identified requirement also serves as the basis for planners to incorporate contractor support into the overall support concept and is communicated through the activity's organizational hierarchy.

STANDARD CONTRACT LANGUAGE

2-57. As contractor support increases in importance to expeditionary operations, it is more and more difficult to rewrite SOWs based on specific operational requirements. Furthermore, a significant effort has been made to capture the variety of contractor deployment, support, force protection, and management requirements from numerous recent operations. Through these lessons, DOD has developed standard contract language that can be used to generically address areas such as deployment/redeployment, accountability/visibility, government-furnished support, and force protection allowing contracting professionals to communicate requirements to contractors. Standard contract language, currently found in the *DOD Acquisition Deskbook Supplement, Contractor Support in the Theater of Operations,* enables contracting professionals to better craft the contracts they award by providing a readily available reference of potential requirements that need to be considered when preparing the contract documents.

CONTRACTOR

2-58. Contractors who have existing contracts with the Army must be included in the planning process as early as possible for a new operation. This ensures that they thoroughly understand the mission and have an opportunity to provide feedback to the military planner on what is commercially feasible and affordable.

Chapter 3

Deploying/Redeploying

INTRODUCTION

3-1. Deployment is the relocation of forces (from both CONUS and outside continental United States [OCONUS] areas) to desired AOs. It encompasses all activities from origin or home station through destination, specifically including intracontinental United States, intertheater and intratheater movement legs, staging, and holding. Redeployment is the transfer of forces and materiel to support another operational requirement, or to return personnel, equipment, and materiel to the home and/or demobilization stations for reintegration and/or out-processing.

3-2. Deployment and redeployment usually pertain to those contractors, either external support or system, that reside outside the AOR and require air or sea transportation to move personnel and equipment to or from the area. Theater support contractors normally reside within such close proximity of the actual AO that the actions associated with deployment and redeployment would not be required. However, when the distances to be traveled are great enough, theater support contractors may be included in the deployment and redeployment process. When such is the case, the governing contract will contain specific provisions regarding the requirements to be followed.

3-3. One of the benefits of using contractor support is that theater support contractors do not usually require deployment support since they are already in the AO. Additionally, in some operations, external support and system contractors can self-deploy using commercial assets. During all contractor deployment/redeployment operations however, the Army establishes measurable performance-oriented requirements and closely monitors contractor performance, including moving personnel and equipment to the AO. Throughout this process, the Army remains responsible for the prudent and legal expenditure of taxpayer dollars even if the contractor self-deploys.

3-4. In operations in restricted environments, the combatant commander often assumes control of lift and port assets. This is especially true when deploying to austere environments and/or limited ports of debarkation (POD) capacity in theater, especially with APODs with limited maximum-on-the-ground capacity. In these operational situations, contractor deployment must be integrated into the combatant commander's deployment priority system. This integration must include establishing Joint Operation Planning and Execution System (JOPES) visibility via the TPFDD process of deploying contractor employees and equipment. The use of habitual relationships between units and their supporting contractors can facilitate the accurate and timely accomplishment of this essential process.

3-5. Responsiveness is a key concern to the Army when determining the use of contractor support. If support is to be effective and responsive, the contractor must be operational at the time it is needed. Some system contractors, especially in digitized units, may be vital to early entry operations and must be placed on the TPFDD accordingly. Support operations that require movement of contractor employees and equipment into an AO must be included in an operations deployment schedule, regardless of whether or not the contractor arranges its own transportation. By doing so, the Army is assured of having the support in place at the proper time.

ROLES AND RESPONSIBILITIES

3-6. Responsibility for ensuring contractor support is available when needed is shared among several commands and agencies, including the contractor. A unity of effort among these activities is paramount in order for contractor support to be properly included in deployment and redeployment operations. The deploying supported combatant commander, the ASCC, ARFOR, and subordinate support commands utilizing contractor support must be aware of and coordinate special contractor deployment and redeployment requirements.

GEOGRAPHICAL COMBATANT/SUBORDINATE UNIFIED/JOINT TASK FORCE COMMANDER

3-7. The supported geographical combatant (and subordinate unified commander, if applicable) is overall responsible for deployment and redeployment operations planned and executed during joint force missions in the AOR. This responsibility includes identification of the movement, timing, and sequence of deploying forces, including contractors in the TPFDD; reception and integration of supporting units and materiel arriving in theater to support the operation; and assisting these units in recovery and reconstitution prior to redeployment. The supported combatant commander validates movement requirements for all forces and agencies deploying or redeploying in support of an operation and provides these validated requirements to USTRANSCOM for planning and execution of strategic movement. At this level major policy decisions regarding the deployment of contractor employees and equipment, including how they will deploy, predeployment processing requirements, and arrival sequencing are first made.

ARMY SERVICE COMPONENT CCOMMAND/ARMY FORCE

3-8. The ASCC commander, in conjunction with the subordinate ARFOR commander(s), is overall responsible for establishing operational specific Army policies and procedures for collecting, verifying, processing, maintaining, and submitting ARFOR deployment planning data. The ASCC is also responsible for ensuring that contractor deployment requirements, based on the combatant commander's decisions, are included in the ARFOR TPFDD planning process and that they receive the priority that enables the contractor to provide support when needed.

SUPPORTING ARMY SERVICE COMPONENT COMMAND/MAJOR ARMY COMMAND

3-9.The supporting ASCCs/MACOMs, typically FORSCOM, USAREUR, USARPAC, and USAMC, are the commands that provide the additional resources to the supported combatant commander. The supporting ASCCs are often responsible for the deployment of their forces in support of other theater requirements, while USAMC routinely deploys some units and/or individuals in support of major, and most minor, Army operations. These supporting Army commands ensure the proper integration of force deployment data into the TPFDD for both units and individuals. The supporting ASCCs/MACOMs must identify any supporting contractor presence and integrate their employees and equipment requirements into the TPFDD and deployment/redeployment structure.

US TRANSPORTATION COMMAND

3-10. USTRANSCOM is the functional combatant command responsible for common-user air, land, and sea transportation, as well as port management for the Department of Defense across the range of military operations. It assists the supported combatant commander to ensure that validated movement requirements, including those for contractors, are routed and scheduled for maximum support during deployment, sustainment, and redeployment.

REQUIRING UNIT OR ACTIVITY

3-11. The requiring unit or activity, as the organization or agency that originates the specific requirement for support, is key to defining deployment requirements for contractor support. Once the decision is made to utilize contractor support, the requiring activity is responsible for identifying when and where the support is needed and for providing the specific data needed to incorporate contractor employees and equipment into the TPFDD. The requiring unit or activity essentially serves as the deployment/redeployment sponsor for the contractor. In the case of ASA(ALT) PEO/PM system contractors, a combination of the requiring unit or activity (the PEO/PM offices) and the supported unit and, if designated, the USAMC LSE is responsible to ensure that the system-contractor employees are prepared to deploy.

SUPPORTING CONTRACTING ACTIVITY

3-12. The various supporting contracting activities, through coordination with the requiring activity, incorporate all contractor-related deployment decisions into appropriate contract language and ensure that all contractor deployment instructions and requirements are included in contracts. These activities may include the contracting offices supporting PEO/PM system managers or those responsible for theater- or external support-type contracts. For contracting activities to accomplish their tasks, the deployment decisions of the commander must be validated as contract requirements and properly funded. A contracting activity cannot enter into a contractual agreement without valid requirements and proper funding. Therefore, planners must communicate their requirements in a timely manner, through the requiring unit or activity to its supporting contracting activity. The supporting contracting activity serves a key role in communicating deployment/redeployment requirements to supporting contractors and obtaining deployment-related information from them and, therefore, must be included in the deployment information chain.

SUPPORTING INSTALLATION/ACTIVITY

3-13. The supporting installation, some of which are specifically designated as force projection (sometimes called power projection) platforms, is responsible for providing support to deploying/redeploying contractors in the same manner as military forces. When directed by the OPLAN/OPORD, these installations assist with the predeployment processing and facilitate the movement of contractor employees and equipment to the designated ports of embarkation (POE). When necessary, they also may be required to assist a deploying contractor to pack, upload equipment, train, and move employees and equipment. Additionally, they may provide assistance or garrison-type life support in staging areas. The terms and conditions of the contract should clearly stipulate the type and amount of this support.

CONTRACTOR

3-14. Contractors share an equal role with government organizations in the deployment/redeployment process. During deployment and redeployment, the contractor is responsible for complying with all related provisions specified in the contract. This includes providing specific employees and equipment data for the TPFDD, complying with prescribed predeployment processing requirements, such as training, medical fitness, and ensuring that their employees are prepared to deploy when notified. The contractor also ensures that its employees are aware of deployment requirements associated with their position, including potential danger, stress, physical hardships, and living conditions.

PREDEPLOYMENT REQUIREMENTS

3-15. Prior to deployment, certain requirements must be satisfied for contractors, their employees, and equipment to be properly prepared to move to begin the deployment process. These predeployment requirements, identified during the planning process and included in the appropriate contract, communicate to the contractor the specifics of how their employees and equipment will flow to the AO. They will prescribe whether or not the military will arrange for and provide the transportation to deploy the contractor, the location deployment will commence from, and the specific prerequisites that contractor employees must meet to travel to the AO.

TRANSPORTATION

3-16. The general government policy is that a contractor provides everything necessary to perform under the contract, including transportation. In many cases, contractor-arranged transportation is the preferred option. However, when the combatant commander assumes control of lift and port assets, contractors may have difficulty obtaining transportation or arrival clearances when deploying. In some operations, it may not be feasible for contractors to self-deploy. Generally, the nature and scale of an operation determines the best method for transporting a contractor to the operational area.

3-17. The influences that help determine the manner in which a contractor is transported are the type of relationship (habitual or nonhabitual) the contractor has with the unit(s) it supports, the type of contractor (system, external support, or theater support), and the point of origin (CONUS or OCONUS). Contractors (typically system contractors) who maintain a habitual relationship with a supported unit should be included in that unit's deployment process and move on whatever government-arranged conveyance the unit's personnel and equipment utilize. Contractors with a nonhabitual relationship and those that travel to the operational area from OCONUS may deploy either by commercially available transportation that they arrange or by government-owned or government-arranged commercial transportation. Whichever method is used must be determined during the planning process and communicated to the contractor through the contract.

3-18. In some recent military operations, the ARFOR commander has designated the USAMC LSE as the lead organization to coordinate all system-contractor employee (to include ASA[ALT] contractors) deployment and to conduct all in-theater system-contractor management. While this is an option to the ARFOR commander, the USAMC LSE is not currently manned nor equipped to execute this mission without augmentation. See Chapter 4 for more information on the roles and responsibilities of the USAMC LSE.

HEALTH, DENTAL, AND PHYSICAL STANDARDS

3-19. The austere environment that external support and system-contractor employees may deploy to and operate in, coupled with the fact that limited medical capabilities exist in the theater, dictates that the combatant commander and subordinate ASCC, through the contract, establish and enforce the requirement for certain health, dental, and physical standards. Although it is not the intent of the Army to dictate to the employing contractor the criteria used to make a hiring decision, the reality of the operational environment requires that standards be imposed on US citizen, and some TCN, external support and system-contractor employees. Therefore, these support contractors deploying to a theater will adhere to established fitness standards and criteria as established by Army and/or ASCC policy. Additionally, the appropriate joint surgeon may articulate additional operational specific medical requirements. All fitness standards and criteria are communicated to the contractor through the terms and conditions of the contract. The contractor ensures that its employees meet these standards, including having all required dental work accomplished prior to reporting to the military deployment force-projection platform so that pre-existing medical conditions do not place an unnecessary burden on the theater medical structure. Employees who fail to meet entrance standards set by the joint and/or ARFOR surgeon, or who become unfit through their own actions (pregnancy, alcohol or drug abuse, etc.) will be removed from the AO at the contractor's expense.

3-20. The final authority for determining fitness is the military physician's evaluation. Figure 3-1 is a representative list of conditions that may medically disqualify a contractor employee from being deployed to an AO.

Individuals with any of these conditions must be evaluated by a military physician to determine if a waiver is appropriate. Under no circumstances should a contractor employee be permitted in the AO without first being medically evaluated.

- Uncontrollable hypertension
- Seizure disorders of any type
- Pregnancy
- Untreated thyroid disorders
- Renal calculus; chronic recurrent
- Raynaud's disease
- Morbid obesity
- Sarcoidosis
- History of chronic kidney disorders
- Diabetes mellitus treated with insulin or oral medications
- Peptic ulcer disease currently being treated or history of surgery for a peptic ulcer
- Chronic low back pain or history of herniated nucleus pulposis
- Any history of cardiac arrhythmias requiring medication
- Any history of myocardial infarction, cardiac surgery or angioplasty with or without stents, without a full diagnostic workup by an attending physician providing clearance for deployment
- Absence of dental panograph on file
- Ongoing treatment of oral pain, trauma, or infection.
- Emphysema or chronic bronchitis
- Asthma
- Any history of transplant surgery
- HIV positive or AIDS
- Any history of heat stroke
- Sleep apnea
- Tuberculosis
- Migrane headaches
- Parkinsons disease; multiple sclerosis
- Untreated symptomatic inguinal hernias
- Cancer, including anyone currently undergoing chemotherapy or recent history of surgery for cancer
- Any neuro-psychiatric disorders, with or without medication
- Anemia of any type until medically cleared by a military physician
- Any condition that may be of such a nature to put the employee at risk medically if deployed
- Any sexually transmitted disease until treated and resolved
- Active orthodontic appliances
- Existence of oral disease that, if not treated, is likely to result in a dental emergency within 12 months.

Figure 3-1. Possible Medically Disqualifying Conditions

3-21. Medical and dental screening of system and external support contractors will be accomplished prior to deploying contractor employees into the AO. This will avoid the costs to the contractor and the government of returning medically unfit individuals from the AO. Contractor employees will be required to present their medical and dental records with a recent history and physical not over 12 months old for screening purposes. The dental record must indicate a dental examination that is not over six months old. If possible, contractor employees should undergo medical screening with their habitually supported unit. Contractor employees who are not screened in conjunction with their supported unit will be individually screened at one of the designated deployment support centers as discussed in the predeployment processing section of this chapter. A recommended health assessment questionnaire is at Appendix C.

Note: at the time of the approval of this FM, HQDA was in the process of updating AR 715-9. This policy update will mandate the use of the health assessment forms for all contractors deploying with the force.

3-22. Contractor employees are normally required to receive the same immunizations as those directed for military personnel. AR 40-562 provides detailed information concerning immunizations. The contractor is responsible for ensuring that its employees are properly immunized prior to deployment to the AO. Governing contracts will provide specific guidance regarding immunizations, including a list of those required for the AO, or instructions on where to go to determine the specific immunizations required. The Centers for Disease Control provide an Internet-based health information service that includes recommended immunizations for each of 17 global regions. Found at www.cdc.gov/travel, this service is frequently used to update information to deploying contractors. Contractor employees must obtain standard immunizations such as tetanus, typhoid, yellow fever, polio, measles, mumps, rubella (MMR), and meningiococcus from their personal physician. The government providew immunizations that are unique to the military and required for entry into the AO.

3-23. The combatant commander or subordinate ASCC commander may also establish limited physical standards for contractor employees when they are placed in a nonstandard work environment, such as those related to military operations. Contractor employees, who would not have difficulty working and living in a normal environment, may have physical difficulties when placed in a field environment. This often austere environment may affect work performance as well as overall health. Physical standards for contractors will not be the same as those expected for military personnel, but should be oriented towards the ability of the individual contractor employee to work and survive in an AO.

PERSONAL READINESS

3-24. Personal readiness involves those actions that the individual contractor employee, with the assistance of the contractor, the supported unit, and force-projection platform, is required to accomplish prior to deployment. These include—

- Obtaining a passport and associated visas.
- Completing required dental work, personal will, powers of attorney, and family care plan.
- Reviewing health and life insurance policies.
- Updating next of kin notification.
- Collecting personal items and clothing.

Contractor employees also need to be advised that personal life insurance coverage may be limited or denied in certain military-related operations. When this is the case, the government is prepared to underwrite the insurance to enable coverage to continue. It is also important to understand that beards, long sideburns, or other facial hair may affect NBC protective mask fit and affect eligibility for deployment. At Figure 3-2 is an individual deployment checklist for personal items.

Individual Deployment Checklist - Personal Items

Passport	Towels and Wash Cloths
Prepare Personal Will*	Shower Shoes
Prepare Power of Attorney*	Underwear (2 weeks)
Prepare Family Care Plan	Socks (ample quantity)
Direct Deposit	Safety Shoes
Review Health Insurance	Sewing Kit
Review Life Insurance	Waterproof Plastic Bags
Blank Checks	Telephone Calling Card
Insect Repellent	Personal Credit Card
Sunscreen	Stationary and Stamps
Lip Balm	Radio (battery powered)
Medical Tags	Hearing Aid Batteries
Extra Pair of Eye Glasses	Pocket Knife
Personal Hygiene Items	Extra Civilian Clothes
Shaving Kit	Extra Batteries
Wet Weather Gear	Flashlight
Personal Medication (90-day supply)	

*Denotes items that SHOULD NOT be physically transported during deployment.

Figure 3-2. Individual Deployment Checklist-Personal Items

TRAINING

3-25. Deploying contractor employees departing from CONUS or OCONUS may require specific training to prepare them for the environment in the AO (Figure 3-3). Training requirements must be identified during planning, specified in the OPORD/OPLAN plan, and stipulated in the contract. The extent and type of the training and equipment will vary depending on the nature of the operation and the type of contractor (theater support, external support, or system) involved. Training may be provided by the military through the supported unit if a habitual relationship exists, through the designated deployment site, or by the contractor himself, utilizing guidelines provided through the contract.

3-26. Training may be provided by the military through the supported unit if a habitual relationship exists, through the designated deployment site, or by the contractor himself, utilizing guidelines provided through the contract.

NOTE: There is no standard mechanism to address the cost associated with this training. The cost issue and other training issues must be worked out on a case-by-case basis.

Training

Law of war/law of armed conflict
Code of conduct
Health and sanitation
Customs and courtesies for the area of deployment
Status of forces agreement
Security
Weapons familiarization
Operational security
Rules of engagement
Chemical, biological, and nuclear protection
Survival, escape, resistance, and evasion
Driver and special equipment training
Other training directed by the combatant commander

Figure 3-3. Potential Training Requirements

3-27. Contractors and their employees often require certain items of GFE (clothing and equipment) to function properly in the AO. These requirements, like training, must be considered and identified during the planning process and included in the contract. Equipment provided as GFE may include—

- Organization clothing and individual equipment (OCIE), including chemical defense equipment (when the threat exists).
- Weapons (for personal defense and only when authorized by the combatant commander, permitted by contractor policy, and agreed to by the individual employee). Weapons can only be a government-issued side arm for personal defense.
- Government-furnished vehicles.

EQUIPPING

3-28. Like training, there are very limited standard policies or procedures for obtaining and issuing GFE to contractor employees. Department of the Army has authorized installations to carry extra CTA 50 items to equip deploying contractor personnel. Supported unit logistics staff may be required to acquire this via operational or pre-positioned stocks, or other sources. In any case, contractors and their employees will maintain and return all issued clothing and equipment back to the place of issue upon return from the AO.

In the event that issued clothing and equipment is lost or damaged, contractors and their employees will be subject to the same requirement for a report of survey, in accordance with AR 735-5, as are government military and civilian personnel. Additionally, the government may require reimbursement from the contractor.

PREDEPLOYMENT PROCESSING

3-29. To ensure that contractor employees and equipment arrive in operational theaters when needed and are capable of operating in a hostile environment, they must be included in predeployment processing. The ASCC has the option of allowing the contractor to accomplish the predeployment requirements on his own or requiring that the contractor process through one of a number of military sites. The choice depends upon the nature of the operation and is identified during the planning process. Either way, the ASCC and its subordinate ARFOR (as applicable) must ensure contractor compliance with the requirements. Specific predeployment processing requirements are communicated to the contractor through the contract.

CONUS REPLACEMENT CENTER/INDIVIDUAL DEPLOYMENT SITE

3-30. The Army has created several sites within CONUS for expeditiously preparing individuals for deployment and to process individuals during redeployment back to CONUS. These sites are known as CONUS replacement centers (CRC) or individual deployment sites (IDS). An IDS is permanently operational during peacetime operations, while a CRC is activated for mobilization only. CRCs and IDSs expeditiously receive and certify individuals for deployment, redeployment, or demobilization; they receive, outprocess, and account for individuals returning from theaters of operations. These individuals are referred to as non-unit-related personnel (NRP) and consist of five categories:

- Active component soldiers (including reserve soldiers accessed onto active duty).
- DACs.
- Contract civilians.
- Red Cross workers.
- Other civilians.

When it is not practical or affordable, contractors deploying from Hawaii or Alaska may not be required to process through either a CRC or IDS. In these cases, predeployment processing will be conducted upon arrival in the theater of operations. In a mature theater such as Europe, this deployment processing may be conducted in a theater processing center similar to a CRC that is in the theater, but outside the actual AO. In worst case scenarios, some contractor employees may have to complete predeployment processing and qualifications in the AO.

3-31. Contractor employees may be required to process through an IDS or CRC. If so, use of these sites is identified during planning and specified, if possible, as a specific requirement in the contract. Although the government conducts the processing, it is still the contractor's responsibility to ensure that its employees are provided all required processing information. As part of the processing, the IDS/CRC may:

- Screen contractor employee records.
- Conduct theater-specific briefings and training.
- Issue theater-specific clothing and individual equipment.
- Verify medical requirements (such as immunizations, DNA screening, HIV testing, and dental examinations).
- Arrange for transportation to the AO.
- Issue DD Form 1173 (uniformed services identification and privilege card), DD Form 489 (Geneva Conventions identify card for persons who accompany the armed forces), and personal identification tags (dog tags).

3-32. To process through a CRC, the appropriate contracting officer or, if authorized, COR issuea letter of authorization/identification, in accordance with paragraph 6005H of the Joint Travel Regulation. (Invitational travel orders, frequently used for government personnel, are not authorized for contractor employees.) This letter authorizes travel to, from, and within the AO, as well as processing at the CRC. It also identifies any additional authorizations and privileges, such as access to the post exchange and commissary, care and treatment at medical and dental facilities, and use of government messing and billeting. The letter of authorization/identification must include the name of the approving government office and the government accounting citation and must state the intended length of assignment in the AO. Contractor TPFDD information is covered in CJCSM 3122.02B—Crisis Action Time-Phased Force and Deployment Data Development and Deployment Execution, Volume III, 25 May 2001, Enclosure G, Appendix A, which can be found at the Joint Electronic Library web site (http://dtic.mil/doctrine/,.mi./.gov links, limited CJCS directives).

3-33. While processing at the CRC, the government will furnish lodging, but will charge a nominal fee for meals. Transportation and travel to the CRC is normally the responsibility of the contractor, with appropriate government reimbursement, as directed by the contract.

FORCE-PROJECTION PLATFORM

3-34. A force-projection platform, sometimes referred to as a power-projection platform, is a designated Army installation that is responsible for assisting in and validating the deployment of Army units moving to an AO. They perform functions similar to CRC/IDS sites, but focus their efforts on units rather than individuals. Contractors deploying to an AO may process through a force-projection platform, either as part of the unit they are supporting (if a habitual relationship exists) or as a separate entity when the number of contractor employees warrants treatment as an organization, such as a LOGCAP contractor.

3-35. The force-projection-platform personnel ensure that the contractor employees, typically those without a habitual relationship and having a larger number of employees, deploying through their installation meet directed predeployment requirements and are properly trained (in accordance with established Army standards) and equipped to operate in the destination AO. Additionally, force-projection-platform personnel assist the contractor in preparing vehicles and equipment for shipment and facilitate their movement to the designated POE IAW the terms and conditions of the contract.

CONTRACTOR-PERFORMED PROCESSING

3-36. A realistic option to government predeployment processing for contractors is for the contractor to perform the contractually mandated requirements. This allows the government to concentrate on its primary responsibility of processing military forces while the contractor assumes full responsibility for preparing its employees to work in the AO. This option may be the preferred option for large contractors, normally external support contractors, deploying a large number of employees and equipment from a number of different locations. However, for contractors who move small numbers of their employees and equipment into the AO, government-performed processing remains the preferred option.

3-37. The decision to allow contractors to perform predeployment processing must be made early in the planning process. This decision must be closely scrutinized and coordinated by the ASCC and represent a clear advantage to the government. When this occurs, the requiring unit or activity provides specific processing requirements which are included in the contract, as communicated in the OPLAN/OPORD. The contract contains specific information (lesson plans and reference material) regarding training to be conducted, types of clothing and equipment required. It will also state the location from where these items and identification-type materials (DD Forms 1173 and 489 and dog tags) can be obtained.

3-38. When contractors provide their own predeployment processing, the Army has less control over the processing and movement. Although it may be less costly for contractors to orchestrate their own deployment, the use of Army deployment centers and transportation by the contractor may save time and resources as well as increase efficiency and control. This concern, however, can be lessened through specifically written contract clauses and an established system of CORs and other contract administrators who would monitor the contractors processing activities.

MEDICAL PROCESSING REQUIREMENTS

3-39. As part of the predeployment process, CRC/IDS medical and dental personnel verify that all requirements for deployment are met. Screening will include HIV testing, pre- and post-deployment evaluations, dental screenings, and tuberculin skin test. Any person who does not meet the established standards will be flagged and then reviewed individually by a military physician or dentist. A determination will be made at that time if the condition in question disqualifies that individual from deploying to the AO. The military physician reviewing the record will consider factors such as age, medical condition, job description, medications, and requirement for follow-up care, and will make the determination for fitness. Medical screening will include completion of DD Form 2766, *Adult Preventive and Chronic Care Flowsheet* (Continuation Sheet), and DA Form 4036-R, *Medical and Dental Preparation for Overseas Movement.* Completed copies of these forms will be included in the individual's deployment packet. Additionally, the CRC/IDS also forwards copies of all DA Forms 8007 to the appropriate surgeon's office.

3-40. Upon arrival in an AO, the contractor provides a summary of each employee's medical history to the supporting medical treatment facility, whether operated by the Army or contractor. This includes the personal physician's (or health maintenance organization) name and phone number, existing prescriptions, immunizations, major underlying conditions (especially those requiring immediate medications or interventions), and other information that needed to provide adequate medical care for the employee. Additionally, for identification purposes, it will be stipulated that contractor employees have a dental panograph and a DNA sample on file.

3-41. For various reasons, some system- and external support contractors may inadvertently deploy without processing through either a force-projection-platform processing center, an IDS, or a CRC. When this occurs, the predeployment processing is conducted in the AO as soon as possible during the reception processing. If the employee is found to have a medical condition that requires them to return to CONUS or their country of origin, it is the employing contractor's responsibility to arrange for the return of the employee and to provide a replacement.

ESTABLISHING ACCOUNTABILITY

3-42. The activities and movement of contractors on the battlefield must be orchestrated along with the combat forces. To accomplish this, commanders must have accountability over all system- and external support contractor employees who deploy with the force, just as with military and DACs. Accountability of contractors in an AO is required for force protection, allocating resources, and administration. Accountability is best established in the pre-deployment stage, capturing the overall deploying contractor employee presence supporting an operation from the point of origin. Chapter 4 discusses contractor accountability in detail.

DEPLOYMENT OF CONTRACTOR EQUIPMENT AND EMPLOYEES

3-43. As discussed earlier, depending on METT-TC, contractors schedule and arrange for their own transportation or utilize government-furnished transportation. This decision is a matter for planners to consider and the commander to determine before an operation begins and, once decided, communicated to the contractor through the contract.

DEPLOYMENT OPTIONS

3-44. The primary factors to be considered when making a determination of the mode of contractor deployment is how restrictive is the theater deployment infrastructure and when the military requires the contractor to be in the AO. When contractors provide their own transportation, the government is relieved from arranging and scheduling the assets needed to move the employees and equipment. Instead, the government simply specifies a time and location for the contractor to provide support and pays the contractor for this service with associated profit and overhead charges.

3-45. Although contractors are generally responsible for providing their own transportation, the government may assist or assume responsibility for deployment transportation. The ASCC G-3, with assistance from the G-4 transportation officer, determines this during the planning process. In many operations, it is expected that US forces will control transportation assets and port facilities. When a contractor does not have full control over the circumstances that affect how it deploys, it may be in the Army's best interest to assist. However, having the military arrange for the transportation may delay the contractor deployment to the AO. In this case, the contractor-arranged option might be preferable. Regardless of the arrangement, the Army must ensure that a contractor's transportation needs are given the same priority as the unit it is supporting, whether the contractor or the Army is responsible for deployment transportation.

Time-Phased Force Deployment Data

3-46. TPFDD is the JOPES data base portion of an operation plan. It contains the time-phased force data, NRP cargo and employees data, and movement data for an operation plan.

3-47. The TPFDD is the combatant commander's statement of his requirements by unit type, time period, and priority of arrival. When considered during the planning process, it includes estimates of contractor cargo and employee requirements along with those of the combat forces supporting an operation. The TPFDD, developed through the process described in Figure 3-4, is both a force requirements and a prioritized transportation movement document. Further, the TPFDD defines the combatant commander's time-phased lift requirements for supplies, equipment, and replacement personnel needed to sustain the forces specified during force planning.

3-48. Responsiveness is key. The combatant commander and his staff ensure that contractor deployment requirements are included in the TPFDD development and that their priority enables the contractor to provide support when needed. Appendix E provides specific instructions for incorporating contractor deployment requirements into the TPFDD.

3-49. Deploying contractors must be included in the TPFDD, whether the government or the contractor provides the actual lift resources. This permits the Army to consciously consider contractor deployment requirements and to incorporate them into the flow of forces in an orderly, scheduled manner. When sequenced with the overall operational plan, these requirements enable the Army to specify when contractors should deploy to the theater and to ensure their timely arrival without disrupting or conflicting with the flow of combat forces.

3-50. The movement data portion of the TPFDD includes—

- In-place units.
- Units deployed to support the OPLAN with a priority indicating the desired sequence for their arrival at the POD.
- Routing of forces to be deployed.

- Estimates of non-unit-related cargo and personnel movements conducted concurrently with the deployment of forces.
- Estimates of transportation requirements filled by common-user lift resources as well as those requiements filled by assigned or attached transportation resources.

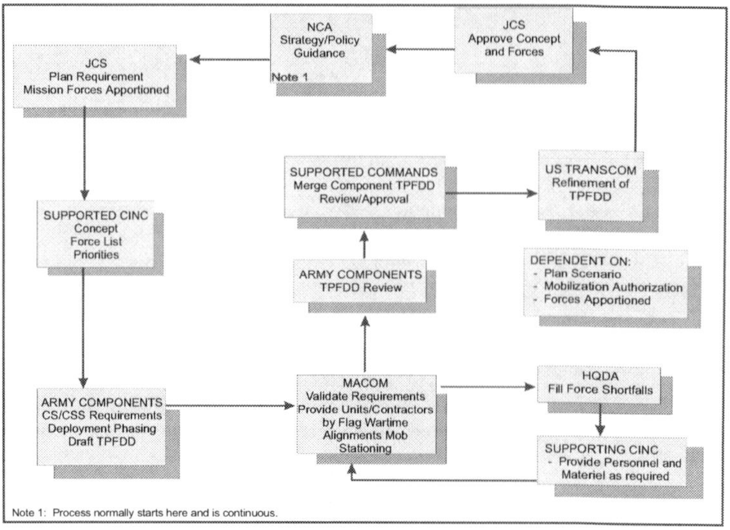

Figure 3-4. The TPFDD Development Process

Contractor-Initiated

3-51. The global nature of contractor support dictates that contractors may deploy employee and equipment from anywhere in the world. Even US-based contractors may have elements originating from locations outside CONUS. Accordingly, contractor-initiated deployment should be the preferred option for deploying employees and equipment, unless circumstances dictate that government control is necessary. By doing so, the government is relieved of the burden of locating lift assets and performing predeployment processing.

3-52. Contractor-initiated deployment is provided for in the governing contract. When permitted, contractors are given appropriate deployment guidance so that they can arrive at the time and location needed without interfering with the deployment of military forces. As a minimum, contracts should provide information to the deploying contractor regarding notification procedures, predeployment processing requirements, and the date and time required to be in the AO. Additionally, the contract should provide for the establishment of management controls that require contractors—

- To ensure that their transportation itineraries follow approved lines of communication and do not conflict with other support and operational traffic. When required, the government should assign mission identifiers and control arrival times.
- To establish and implement in-transit visibility procedures, to include listing the Air Mobility Command Tanker Airlift Control Center (TACC) as an information addressee on position reports submitted through Aeronautical Radio Inc. and Societe International de Telecommunications Aeronautiques.
- To establish a means for the supported commander's command and control center to immediately contact the operations center of the transportation provider so that emergency recall or diversion instructions can by issued to contractors who are en route on air, sea, or land transportation.

PORTS OF EMBARKATION

3-53. Designating a POE from which personnel and equipment depart facilitates the movement of contractor employees and equipment according to the established priority of the combatant commander and subordinate ASCC. These activities facilitate contractor movement, thus assisting with preparation and loading of cargo and employees in the same manner they do for deploying military forces. However, port managers cannot properly incorporate them into their deployment operations unless they are aware of their responsibility. Planning and TPFDD must designate POEs that support deploying contractors and inform the required personnel of their support responsibilities.

3-54. For deployments under government control, Military Traffic Management Command (MTMC) port-call instructions direct the flow of contractor employees and equipment into the POE. Once at the POE, contractors follow the processing instructions issued by the responsible management activity. For seaports of embarkation, MTMC is the worldwide common-user ocean-terminal port manager. It directs deployment of units and sustainment according to the TPFDD. For aerial ports of embarkation, aerial port squadrons and tanker airlift control elements, provided by the Air Mobility Command, furnish the airlift interface. In conjunction with the departure airfield control group and designated Army activities, contractor employees and equipment travelling by air are processed for deployment. At the POE, contractor employees and equipment are handled in the same manner as deploying military forces and receive the same level of support as military units.

ARRIVAL IN THEATER

3-55. Contractor employees and equipment must flow into the AO in a controlled and managed fashion, usually in accordance with the TPFDD. This provides the combatant commander and subordinate ASCC with visibility over incoming contractor employees and cargo, affords the opportunity to deconflict the use of air and sea ports, and avoids placing contractors at risk of becoming targets as unidentified inbound aircraft or ships. Contractor arrival is handled in the same manner as with arriving military forces.

PORTS OF DEBARKATION

3-56. The PODs, either aerial or sea, are the geographic points at which cargo and personnel are discharged. APODs serve as the primary port of entry for all deploying personnel, as well as for early entry forces normally airlifted into an AO with their equipment. Activities at the APOD include terminal control; loading, unloading, and servicing of aircraft; clearing of personnel and cargo; and life support. Seaports of debarkation (SPOD) serve as the primary ports of entry for equipment, where vessels are off-loaded, and where cargo and personnel are cleared and moved to the staging area for movement forward. Contractor employees and equipment, either through the TPFDD or instructions contained in the governing contract, process through PODs in the same manner as military forces.

RECEPTION AND INTEGRATION

3-57. Effective force projection largely depends on the speed with which forces assemble at the required location. The essential process that transitions deploying forces into forces capable of meeting the combatant commander's operational requirements is reception, staging, onward movement, and integration (RSO&I). JP 4-01.8 and FM 4-01.8 (100-17-3) contain detailed discussions of the RSO&I process. All contractors (especially external support, system, and, to a limited extent, theater support) must be included in the RSO&I process to perform in harmony with other support capabilities. Contractor employees and equipment must be received and integrated into the operational support structure, whether they originated in CONUS, a third country, or the host nation. Contractor reception processing establishes in-theater accountability of contractor employees. A formal reception and integration process is essential for both system- and external support contractors deploying into the theater. For contractors who reside within the AO or travel only a short distance, reception and integration is less formal. Reception of contractors begins upon arrival in the AO at the APOD/SPOD. When arriving by government transportation, personnel and equipment are processed in the same fashion as military forces. For those arriving on carriers provided by the contractor or for contractor employees that are already in the AO, reception processing is similar but requires the contractor to be integrated into the processing flow. Regardless of the situation, the requirement for reception processing must be addressed in the governing contract.

3-58. Reception is the process of unloading personnel, equipment, and materiel from strategic or operational transport and marshalling the deploying units or activities for movement out of the reception area. During major deployments, the majority of personnel arrive via strategic airlift and most equipment and materiel arrive by strategic sealift.

3-59. Reception operations include all of those functions necessary to receive and clear personnel, equipment, and materiel through the APOD/SPOD. Upon arrival, external support and system contractors flow through the same processing steps as their military counterparts. Personnel and cargo are off-loaded at terminals, processing through the reception center to determine how and where they are to be moved. During reception, accountability is continued for contractors originating from established military POEs or established for those contractors entering the military system for the first time. Once initial processing is completed, contractor employees and equipment are integrated into the military movement control system for movement to their designated destinations.

3-60. System contractors often have a habitual relationship with the unit they support. These contractor employees deploy with and go through the RSO&I process with their supported military unit. Area support-system contractors deploy with a support unit as designated by the ASCC or subordinate commander. External support contractors without a habitual relationship undergo the same RSO&I process as military forces.

3-61. Although theater support contractors do not have to deploy to the AO, they still must undergo a reception and integration process and may require tailored, theater-specific training and/or other RSO&I-related actions. They must be integrated with the unit or activity they will work for and may require issued locally procured ID cards if they require access to military-controlled areas/facilities. Theater support contractor reception and integration processing, as directed by their contract, is provided at an AO reception center by the activity contracting for the support or from the unit/activity receiving the support. However this processing is accomplished, these contractors must be processed and integrated into the support structure.

3-62. Once reception is completed, military forces proceed to the next phase of RSO&I—staging and onward movement. Staging includes the assembly, temporary holding, and organizing of arriving personnel, equipment, and materiel into units and forces and preparing them for onward movement and employment. Contractors usually do not proceed to the staging phase, but will be directly integrated into the support structure, which may include limited onward movement to their final destination. Contractors may proceed to the staging phase when they accompany a deploying military unit, or they are awaiting additional employees or cargo before they can proceed to their destination. In this case, contractors follow the same procedures as those for military forces. If staging and/or onward movement of a contractor is required, military planners must recognize the requirement so that the support structure is prepared to handle the additional workload. Additionally, governing contracts will contain appropriate language directing the contractor to adhere to the staging requirement.

REDEPLOYMENT

3-63. Upon completion of an operation, contractors redeploy out of the AO as quickly as METT-TC will allow. The timing of the departure of contractor support operations is as critical as that for military forces. Orderly withdrawal or termination of contractor operations ensures that essential contractor support remains until no longer needed and that movement of contractor equipment and employees does not inadvertently hinder the overall redeployment process. FM 3-35.5 (100-17-5) describes in detail redeployment and the procedures involved in redeploying military forces. To an extent, the same procedures apply to redeploying contractors. However, planners must determine the specific steps desired and be aware of the cost associated with doing so. Because contractors are being paid to perform the support required, it should be terminated when no longer needed. Therefore, the cost effectiveness of requiring a contractor to redeploy through the same stepped sequence as military forces must be assessed before arbitrarily requiring a contractor to do so.

TRANSITIONING SUPPORT

3-64. As the operational situation dictates, contractors may redeploy prior to the completion of the operation. Quality and responsiveness of support, force protection, cost, or transfer to civilian control are some of the reasons that may create this situation. When contractors redeploy and the support they are providing is still required, there must be an orderly transfer or transition of that support. Support may be transferred to a military unit, another contractor, or the host-nation infrastructure. Regardless, the transitioning of support must be considered and planned for, so that the redeployment of a contractor does not result in the loss of support.

PORT OF EMBARKATION

3-65. Redeployment POEs perform essentially the same functions as those involved in deployment. The procedures are similar, whether contractors are redeploying to their point of origin (home station) or to another AO. Redeploying contractors will normally do so in the same manner in which they deployed (either under government control or self-deploy).

3-66. Prior to arriving at the aerial ports of embarkation/seaports of embarkation, contractors have accomplished the same preparations as the military forces. Depending on the destination, specific wash-down and customs requirements for vehicles and equipment are met; vehicles and equipment are prepared for shipment; and any GFE issued while in theater is returned. Additionally, transportation arrangements are made for the contractor's employees and equipment. If redeploying to another AO, contractors may be provided with equipment and supplies and any additional training required to conduct follow-on support. Once all preparations are completed, contractor employees and equipment proceed to the designated port (air or sea) for final processing and departure. Similar to deployment, contractor accountability measures continue, enabling the military to maintain accountability of and manage contractors as they proceed through the redeployment process.

CONUS RECEPTION AND OUTPROCESSING

3-67. Contractor employees who deployed through the CRC or IDS should be required to return through the same processing center for final outprocessing. The CRC/IDS is responsible for assisting the return of individual contractor employees and ensuring employee protection, privacy, and transition from the deployment area to home. Contractor employees that deployed with their habitually supported unit normally redeploys with that unit. At the return processing center, contractor employees are required to return any issued clothing and equipment. They receive a post-deployment medical screening and briefings on signs and symptoms of diseases to watch for, such as tuberculosis. Lost, damaged, or destroyed clothing and equipment are subject to a report of survey IAW AR 735--5 and may result in reimbursement from the individual's employer. Normally, the amount of time spent at the return processing center will be the minimum required to complete the necessary administrative procedures. Transportation for contractor employees from the return processing center to their home destination is normally the employer's responsibility. The provisions of the governing contract determines government reimbursement to the contractor for the travel.

Chapter 4

Managing

INTRODUCTION

4-1. Effective contractor management on the battlefield is essential to ensure that contractor-provided support is properly orchestrated and synchronized with the overall operation support plan, and that contractor employees are properly accounted for, protected, and supported. Additionally, adequate contractor-employee accountability and contractor visibility in the theater is necessary to establish positive control, to perform initial reception and integration, to provide necessary support, and to establish and manage their location and movement on the battlefield. Appendix B provides a notional example of contractor management planning requirements.

4-2. As stated earlier, contractor management does not flow through the standard Army chain of command. Management of contractor activities is accomplished through the responsible requiring unit or activity COR through the supporting contracting organization in coordination with selected ARFOR commands and staffs. It must be clearly understood that commanders do not have direct control over contractor employees (**contractor employees are not government employees**); only contractors directly manage and supervise their employees. Commanders manage contractors through the contracting officer and their appointed CORs in accordance with the terms and conditions of the contract.

ROLES AND RESPONSIBILITIES

4-3. Commanders and their staffs at all levels must manage the total contractor force supporting the ARFOR units within the AO. In planning and execution, direct command interest and involvement is necessary to ensure that contractor support receives the proper attention. Contractor support must be monitored to ensure compliance with the stated requirements so it does not hinder the overall military operation.

GEOGRAPHICAL COMBATANT/SUBORDINATE UNIFIED/JOINT TASK FORCE COMMANDER

4-4. The combatant commander sets the tone for the use of contractor support by ensuring it is considered during planning and is part of the OPLAN or OPORD. The combatant commander establishes priorities, determines guidance on contractor use to include specific theater-specific entrance requirements, and restrictions and policies. The combatant commander also ensures policy is in place for the staff, subordinate joint force commands, normally in the form of a JTF, and Service component commands to account for the total contractor presence in the theater to include individual accountability as required, and reviews and approves Service component support plans. The JTF commander is responsible for operations within the joint operational areas (JOA) and, when necessary, may establish JOA-specific contractor policies and procedures. Normally, the combatant commander appoints a lead Service or agency responsible for JTF contractor-related planning and to coordinate common user logistics (CUL)-related theater support contracts. In many cases, the Army will the lead Service responsible for these CUL theater supprt contracts.

ARMY SERVICE COMPONENT COMMAND/ARMY FORCES

4-5. The ASCC commander commands and controls all assigned and attached Army forces in the supported combatant commander's AOR and is responsible for all Army support, including contractors, unless the combatant commander directs otherwise. In most circumstances, the ASA(ALT) designates the ASCC as the HCA, thus making the ASCC the lead contracting authority for the Army in the AOR. The ASCC, through the PARC, manages and conducts Army in-theater-related acquisition activities necessary to accomplish the mission; acquires theater support contract supplies, equipment, and services; provides overall guidance concerning Army theater support contracting plans/policies.

4-6. The ASCC commander is also responsible for the overall management of all contractors providing support to ARFOR units within the AOR. However, unlike the situation for contracting authority, there is no lead ASCC contractor management staff. Without a lead contractor management staff, it is imperative that each staff be familiar with and properly execute their specific contractor management responsibilities as laid out in this chapter.

4-7. As stated above, the combatant commander may also appoint the Army as the lead Service for contracting within a specific JOA. Normally, this lead Service mission is executed by a subordinate ARFOR contracting agency, called the mission chief of contracting in joint vernacular, which acts as the approving authority for all common-user contracting actions for the joint force as designated by the combatant commander. The ASCC and ARFOR in this case would also be responsible to develop the joint contracting support plan. See JP 4-0 and JTTP 4-07 for more information on contracting in joint operations.

PRINCIPAL ASSISTANT RESPONSIBLE FOR CONTRACTING

4-8. A special staff officer is the ASCC senior Army acquisition advisor responsible for overall **planning and management of <u>theater support</u> contracts in the AOR**. The PARC is often dual-hatted to command the Army's theater contracting activity and to oversee the daily activities of the ASCC theater support contracting operations. With regards to management of contractors in an AOR, the PARC is the focal point in the planning for and managing of those contracts let by the Army contracting staffs within the AOR. The PARC office is not designed nor manned to function as the lead organization for the overall contractor integration mission; this is a collective function shared by numerous ASCC staff members.

PROGRAM EXECUTIVE OFFICER/PROGRAM MANAGER

4-9. ASA(ALT) PEOs or PMs are the designated managers for the development, acquisition, and fielding of major weapon and equipment systems for the military (less simulations and training devices which are still managed by USAMC). The PEOs and PMs are responsible for planning, programming, budgeting, and execution needed to place a particular system in the hands of the ultimate user. During the fielding of a system, some governing life-cycle contracts are written to include contractor-provided support after the system has been sent to the user either as ICS (temporary contractor support) or CLS (long-term contractor support).

4-10. Whether a system is ICS or CLS, PEOs and PMs include deployment provisions into their contracts so that the system contractor involved is aware of and prepared to live and operate in possibly austere and hazardous environments. Important among these requirements are the management and accountability procedures used when the contractor is in the AO. This requirement is often difficult to manage, especially since most PEO and PM management personnel do not deploy with the supported unit. PEOs and PMs must establish deployment preparation and in-theater management agreements with the supported field units.

> **Managing ASA(ALT) System Contractors**
>
> At the time of the approval of this manual, the general, but not formal, agreement was that many of the ASA(ALT) system contractors will be managed at the unit level via the USAMC LSE. Accordingly, the PEOs/PMs, along with the USAMC LSE and the supported unit, must ensure that all deploying system contractors are properly prepared for deployment, as well as accounted for, protected, and supported upon deployment.

ARMY FORCE CONTRACTING ORGANIZATION

4-11. The senior Army contracting staff provides Army contingency contracting support in the AO. This organization performs the contracting functions for the ARFOR and ensures all approved requests for theater contractor support are processed IAW joint and ARFOR command guidance. The senior ARFOR contingency contracting organization may be a separately established organization or come from one of the existing ASCC contracting commands. It also could be established utilizing contracting staff found in the theater support commands (TSC), corps support commands, division support commands, or interim brigade combat teams (IBCT).

4-12. The ARFOR contingency contracting organization implements the JTF and ARFOR contracting guidance found in theater contracting support plans. It also negotiates, awards, administers, and closes theater support contracts. The ARFOR contingency contracting organization maintains visibility over their own theater support contractors and the compliance of these contracts, but does not concern itself with the total contractor presence in the AO.

SUPPORT COMMANDS

4-13. The Army maintains a selected number of operational-level support commands. These support commands include the personnel command (PERSCOM), theater-level military intelligence brigades, TSC (logistics), medical command, engineer command, and the theater signal command.

4-14. These support commands are the primary requiring organizations to maintain the day-to-day management, as well as contractor visibility and contractor-employee accountability within their functional area. In small-scale contingencies where these commands may not deploy, the senior support commands performing these operational-level functions would be responsible for the same contractor-support oversight. This mission, depending on the scope of the requirement, may require augmentation from the higher-level support command.

4-15. Some strategic-level commands such as USAMC, USTRANSCOM, DLA, USACE may contract specific support in an AO. When they do so, they, of course, have a responsibility for managing contractors. This responsibility starts with the inclusion of theater-specific contractor-management and visibility requirements in the applicable contract language. Obtained from operational planning documents, these contracts include the various deployment, reception, integration, and reporting requirements necessary to properly manage and control the flow into and operating in the AO. In some cases, these commands are integrated into the Army operational-level support commands addressed above. In other cases, these commands may report directly to the joint force commander. For example, the USAMC LSE, which often plays a vital role in the area of systems-support contract supervision, is normally attached to the TSC or, when the TSC is not deployed, the senior multifunctional logistics command in the AO.

4-16. These strategic-level organizations retain contracting authority for those specific functions from their parent commands. When appropriate, Army commanders and their staffs include these commands in their mission planning, and each should include support appendices in the applicable staff section annex to the OPLAN or OPORD. For example, the staff engineer coordinates USACE procurement of real estate and real property; the joint force transportation planner coordinates with USTRANSCOM commands to monitor their assets. When procuring goods and services within the theater, they will do so IAW guidance and procedures provided by the joint force commander's mission chief of contracting.

4-17. In addition to the strategic organizations listed above, DCMA provides contract-management services to major contracts to include selected contracts within the AOR. DCMA monitors contractors' performance and management systems to ensure that cost, product performance, and delivery schedules are in compliance with the terms and conditions of the contracts.

4-18. The combatant commander J-1 establishes and publishes basic theater entrance requirements, accountability, and other personnel-related procedures to include contractor employees. The combatant command or subordinate JTF command J-1 also coordinates and publishes a specific theater/JOA contractor identification (ID) card policy. The senior Army personnel planner in theater, via guidance from the combatant commander J-

1 and DA PERSCOM, establishes Army-specific contractor-employee theater-entrance, personnel-related support and accountability policies.

CONTRACTOR

4-19. The contractor implements and complies with all government-directed management requirements contained in the contract. Although the contractor is solely responsible to manage its employees and operations, it also has responsibility for ensuring that the various contractually mandated requirements are met. These may include—

- Providing feeder data for the development of the TPFDD and the theater personnel accountability system.
- Ensuring all employees comply with DOD, Army, or theater-specific policies and directives.
- Dealing with performance or conduct problems reported by the government.

CONTRACTOR VISIBILITY AND CONTRACTOR-EMPLOYEE ACCOUNTABILITY

4-20. The intent of contractor visibility and contractor-employee accountability is to know which contractors are participating in an operation, where they are in the operational flow (deployment, reception, integration, or providing support), when they are needed to provide support, and a clear day-to-day overview of their support mission. The ARFOR commander should maintain visibility over all contractors in the AO, similar to the manner we report military unit readiness. This is necessary to orchestrate their activities and movements with combat forces, and so that their relationship with the current operational situation can be assessed and adjusted when necessary.

4-21. Contractor-employee accountability is similar, but not the same, as for military and DACs. Contractor-employee accountability is limited to US citizens and TCN system and external support contractors who deploy with, and are in direct support of, ARFOR units within the AO. Additionally, the combatant, JTF, and/or ASCC commander may expand this individual accountability requirement to include theater support contractors who require access to military facilities. Establishing contractor visibility and contractor-employee accountability requirements by type of contractor is based on the following rationale:

- Theater support Contractors. Normally, theater support contractors are drawn from the local vendor base with their employees usually residing in local communities. As a result, they do not normally require government-furnished life support when supporting an operation. In fact, may theater support contractor employees will not even have access to military facilities and even those that do will often return to their homes or other non-military-supported billeting arrangements. Therefore, detailed contractor-employee data may not be required and only aggregate data indicating the size of the contractor's workforce is collected. However, when security measures require employees to be issued ID cards for entry into a military-controlled area or facility, at a minimum, the employee standard name line and identification number may be required, but would not be reported through the personnel accountability system, in order to issue requisite ID cards.

- External support and System Contractors. External support and system contractors normally are drawn from outside the AO and must travel from either CONUS or OCONUS locations; therefore, they require by-name accountability to support the deployment and redeployment process, as well as support operations. Furthermore, because of inherent responsibilities to its citizens, the US Government provides certain support that it is not obligated to do for other nationalities. In this context, the support relates to the injury or death of a US citizen employee and involves certain rights and privileges associated with notification and recovery, return, and burial of remains. In these cases, specific contractor-employee data is required and reported through the established personnel accountability system. Basic employee data is required for TCN contractor employees who deploy with the force for deployment life support and security reasons.

VISIBILITY AND ACCOUNTABILITY POLICY

4-22. Currently, there are no standard joint or Armywide deployed contractor visibility nor contractor-employee accountability procedures outside of this manual. However, contractor visibility and contractor-employee accountability is needed to ensure that the overall contractor presence in a theater is synchronized with the combat forces being supported. Additionally, the Army's responsibility to provide operational and life support and force protection to (and sometimes from) participating contractors dictates the need for maintaining visibility, and selected accountability, of all contract and contractor employees.

4-23. Maintaining contractor visibility and contractor-employee accountability is a continuous process conducted throughout the duration of the operation beginning at its point of origin. Once initial management of the deployed contractors is established and they are integrated with the unit/activity they are providing support or working for, visibility is maintained through status reports flowing through normal military channels. Governing contracts must stipulate the specific visibility reporting requirements, including the type of information required and to whom the information is to be provided. At a minimum, standard contract language should be used to identify the requirement to provide visibility and accountability reports required.

CONTRACTOR EMPLOYEE ACCOUNTABILITY

4-24. The senior Army personnel planner in theater, via guidance from DA PERSCOM and the combatant commander, establishes operational specific contractor-employee accountability policies. Contractor-employee accountability normally includes maintaining sufficient data to properly plan for and execute medical, casualty, mail, and other support and force protection requirements. At a minimum, the Army personnel system accounts for US citizen contractor employees and TCNs who deploy with the force. Normally, by-name accountability of TCN external support contractors who do not deploy through military means and theater support contractor employees is not required. However, the senior Army personnel organization may be required to ensure that theater support contractor employees are issued theater-specific ID cards, as required by the joint force or ARFOR commander.

ACCOUNTABILITY REPORTING SYSTEM AND REQUIREMENTS

4-25. The senior Army personnel organization in the AO establishes contractor-employee accountability through the requiring unit or activity via TPS. The supporting personnel unit at home station, at the CRC/IDS, or in the AO issues contractor employees civilian ID cards IAW AR 600-8-14. Contractor-employee accountability data is entered and reported via by the requiring organization or activity via TPS IAW operational-specific guidance. Typical contractor-employee accountability requirements include:

- Employee standard name line.
- Point of contact (COR) information.
- Employee sex.
- Employee date of birth.
- Employee religion.
- Employee identification number (social security or other uniquely identifying number).
- Employee passport number (if applicable).
- Employee nationality and home address.
- Blood type.
- Marital status.
- Employee point of departure.
- Deployment date (employment date for theater support contractors).
- Security clearance level.

CONTRACTOR VISIBILITY

4-26. All requiring units are responsible for maintaining visibility for contractors providing support to Army forces in the AOR IAW ASCC OPLAN/OPORD guidance. This visibility data is forwarded through the normal chain of command to the ARFOR staff responsible for oversight of that particular contractor function. The ARFOR staff provides contractor visibility information to the ASCC commander/staff as required.

CONTRACTOR VISIBILITY REPORTING SYSTEM AND REQUIREMENTS

4-27. Currently, there is no approved STAMIS or C^2 system for reporting contractor visibility nor formal policy that lays out the required data fields. Because of this shortfall, the specific contract visibility reporting system and supporting information fields are determined IAW the combatant commander and ASCC guidance. Based on recent lessons learned, any commonly used office automation software can easily accommodate the requirement. Typical contractor and contractor employee required data should include:

- Contract number.
- Contractor company name.
- Type of support provided (e.g., Apache maintenance, road construction, dining facility operation).
- Name of contractor's in-theater manager.
- Awarding contracting office.
- Contracting officer contact information (include name, organization, location, and phone number).
- COR contact information (include name, unit identification code (UIC), and phone number).
- Supported UIC.
- Contractor workforce composition (by location and nationality)—
  - Total number US citizen employees.
  - Total number TCN employees (specific nationality).
  - Total number local national employees.

REQUIRING ORGANIZATION RESPONSIBILITIES

4-28. Requiring organizations, such as the USAMC LSE, USACE, and the ARFOR support commands, must ensure that they provide contractor visibility reports and contractor-employee accountability reports as prescribed in the OPLAN. These commands also coordinate the contractor visibility and employee reporting requirements with the appropriate contracting activity to ensure that this requirement is clearly addressed in the appropriate SOW.

4-29. The contractor maintains data on its workforce and reports required data as prescribed in the contract. The COR ensures that all reporting requirements are submitted to the requiring unit or activity IAW their SOW.

CONTROLS

4-30. The success and effectiveness of the contractor visibility and contractor-employee accountability system are determined by the controls established. Formalizing the requirement during the planning process and incorporating it into the governing contracts legalize the requirement for contractors to provide the needed data. To ensure compliance, a pass and identification system must be in place. Standard civilian ID cards (DD Form 1173) or the new *smart card* ID card must be issued to all supporting system and external support contractor employees at the point initial accountability is established, either at the force projection deployment site, a CONUS CRC/IDS, or the AO reception center. Theater support contractor employees who require access to military-controlled facilities are normally issued theater/JOA-specific ID cards at the AO reception center.

4-31. If this ID card process is not accomplished, contractor employees will not be allowed to enter work areas or utilize government-furnished life support facilities and services, thus placing the contractor, and possibly the Army, in the position of noncompliance with the terms and conditions of the contract.

ESTABLISHING IN-THEATER MANGEMENT OF CONTRACTORS

4-32. Positive in-theater management of contractors begins with the proper integration of contractors into the operational flow. This process is accomplished for the same reason as for military forces, but also because it is the key point for gaining positive management control over contractors preparing to support the military forces. Integration of contractors is accomplished after they are received into the AO. It is not necessarily a distinct phase, as it is with military forces, as much as it is a set of actions necessary to properly include the contractor into the operational support structure. This means getting them into position or moving them to a location to perform the work they are hired to perform. It also involves formalizing contact with administering contracting and supported unit representatives and establishing support relationships.

4-33. The goal of integrating contractors is to ensure that visibility and applicable accountability are established, and they are incorporated into the operational support structure without hindering or disrupting the military forces they are to support. The method in which a contractor is integrated into the theater's operational support structure is driven by the point of origin of the contractor, either CONUS-based, third country-based, or locally based. Regardless of the point of origin, all contractors must be integrated into the operational support structure.

NOTE: All contractors must be processed and integrated into the ARFOR support structure.

CONUS-BASED CONTRACTORS

4-34. The deployment process facilitates integration of CONUS-based contractors. Whether they are deploying under the control of the military or through their own arrangements, these contractors are required, by the terms and conditions of the contract, to process through an in-theater personnel reception center either as a part of their supported unit or individually. If deploying individually, once reception processing is completed, the contractor is linked-up with the AO activity that it will be providing support for, if it did not deploy with that activity. This requiring organization or activity is not only the contractor's customer, with responsibility for directing its requirements according to the terms and conditions of the contract, but also its sponsor, responsible for ensuring that visibility and accountability are maintained and that government-provided equipment and services, including life support, are provided. Integration is complete when visibility is verified or established, and the contractor is ready to provide support.

THIRD-COUNTRY-NATIONALCONTRACTORS

4-35. Third-country-based US or TCN contractor employees should deploy through an OCONUS deployment site if at all possible. In some situations, processing third-country-based contractor employees may take place within the AO. Once processed, these third-country-based contractors are integrated into the operational support structure in the same manner as CONUS-based contractors. They will join their supported unit, where visibility is verified, and then proceed to the location(s) where support will be provided. The key to integrating third-country-based contractors is to ensure that the governing contract directs them to process through the AO reception center so that positive management control can be established.

LOCAL CONTRACTORS

4-36. Local theater support contractors must be properly integrated into the operational support structure. Because theater support contractors do not deploy to the AO, they do process through the reception center, with the exception of getting ID cards as required to meet force protection needs. Therefore, their integration is an abbreviated process involving them furnishing specified information concerning who they are, what support they provide, where the support will be provided, and the composition and size of their workforce, to the sponsoring activity so that visibility can be established. Once accomplished, theater support contractors are assimilated into the operational support structure and are managed in the same manner as any other type contractor.

LOCATION AND MOVEMENT ON THE BATTLEFIELD

4-37. Army operations may occur in a non-linear operational environment without clearly defined traditional borders or boundaries. In these circumstances, contractors can expect to perform virtually anywhere in the AO, subject to the terms of the contract and the combatant commander's risk assessment.

4-38. Army policy (AR 715-9) states civilian contractors may be employed anywhere in the AO necessary to support operations and weapon systems. Generally, contractors are assigned at eachelons above division (EAD); however, if the ARFOR commander, based on combatant and JTF commander guidance and METT-TC and the terms and conditions of the contract, determines that contractor services are required at lower echelons, they may be temporarily deployed into the division area as needed, consistent with the terms of the contract and the tactical situation. In cases where critical system contractors are required forward on a permanent basis, the ARFOR commander requests an exception to policy from HQDA. In reaching his decision, the ARFOR commander considers joint and ASCC guidance, the risk to the mission, and the risk to the contractor employees.

4-39. Contracts for contractor support must be carefully drafted to specify the services needed and the conditions under which they are required so contractors are fully aware of what is involved. When contractors choose to perform under dangerous conditions, the cost of the contract may be increased due to the risk and additional difficulty the contractor is being asked to accept. Contractors may be more likely to perform under dangerous conditions if the Army meets certain security requirements to ensure their protection and safety.

4-40. The portion of a theater adjacent to the APOD/SPOD contains the organizations, lines of communications (LOC), and the theater logistics bases required for immediate support and sustainment of forces in the field. Because this area contains the theater logistics bases, it is where the majority of contractor support operations are found. Within this area, contractor support operations are located where they are the most responsive. Contractors must be located so their operations do not adversely impact the mission or the safety and security of their own employees.

4-41. When addressing where contractors should be located within the AO, planners must also consider the need for contractors to gain access to LOCs and transportation assets to perform their mission. In doing so, they must rely on either the military resources or allow contractors to move about using their own assets. When visibility of contractor-provided assets (e.g., supplies) is necessary, additional management challenges will arise. When contractors are required to use the in-transit visibility/automated information technology devices to provide distribution management visibility, they must either be furnished as GFE or be of a type that is able to interface with the DOD format. In satisfying this requirement, the military may decide to include contractor transportation needs with those of its own and provide transportation support as a government-furnished service. The benefit is a reduction of cost and roadway congestion. The risk is that the contractors' needs may not receive the priority they require, causing them to fail to meet their support requirements. On the other hand, when contractors provide their own assets, the military must ensure that any movement along LOCs, or elsewhere in the theater, is controlled, protected, and does not interfere with other operational requirements. Additionally, consideration must be given to the practicality of military movement restrictions and convoy requirements and their applicability to contractor movements. Is a contractor required to comply with or excluded from the requirements? A conscious decision must be made regarding this issue because of the impact on contractor responsiveness and military force structure impacts (i.e., having to provide personnel to a contractor to comply with the two-driver rule).

CONTRACT COMPLIANCE

4-42. One of the key elements in the managing of contractors is contract compliance. Contract compliance is simply ensuring that the contractor is doing what the contract requires. The key to effective contract compliance is making sure that planners consider the variety of requirements relating to contractor support, include them in operational plans, and communicate these plans to the contracting structure so that they can be included in applicable contracts.

4-43. Contract compliance is the responsibility of the contracting structure that supports the AO. This structure includes the awarding contracting activity (examples include the theater contracting command, USAMC, DLA, or USACE) and the appointed COR. Depending on the scope of the contract and the awarding contracting activity, contracts may be passed to the DCMA for contract administration and compliance. When this is the case, the awarding contracting activity passes routine contract administration responsibilities to the on--the-ground DCMA activity, with the parameters of the delegated responsibility being specified. Regardless of how contract administration is accomplished, the responsible activity or individual maintains insight into the contractors' processes to ensure that the product or service, cost, and schedules are in compliance with the terms and conditions of the contract and theater-specific requirements are being met. Contract compliance includes on-site surveillance and program-specific processes that cannot be monitored by off-site contracting agencies.

LEGAL CONSIDERATIONS

4-44. When managing contractors on the battlefield, military managers, including commanders, must be aware of the legal considerations that influence their management approach. There exists a body of laws, statutes, and regulations that govern the manner and extent that contractors support a military operation and guide contractor support on the battlefield. These legal provisions impact on contractor status and security, contractual provisions, and discipline. The legal authorities affecting contract support include the SOW for the specific operations, international agreements affecting the theater of operations, and general international law (e.g., the Law of War). These provisions influence the manner in which contractors are managed and may limit the extent of a commander's discretionary authority over contractor requirements and management.

DISCIPLINE AND THE COMMANDER'S AUTHORITY

4-45. Contractor employees are not subject to military law under the UCMJ when accompanying US forces, except during a declared war. Maintaining discipline of contractor employees is the responsibility of the contractor's management structure, not the military chain of command. The contractor, through company policies, has the most immediate influence in dealing with infractions involving its employees. It is the contractor who must take direct responsibility and action for his employee's conduct.

4-46. When criminal activity is involved, international agreements and the host-nation's laws take precedence. In the absence of any host-nation involvement, the commander may be able to utilize a recently enacted law to deal with felonies. In November 2000, the Military Extraterritorial Jurisdiction Act (Public Law 106-778) of 2000 was passed by Congress and signed into law. This law permits the prosecution in federal court of civilians who, while employed by or accompanying the armed forces overseas, commit certain crimes. Generally, the crimes covered are any federal-level criminal offense punishable by imprisonment for more than one year. The law applies to any DOD contractor or subcontractor (at any tier) or their employees provided they are not a national of or reside in the host nation. The law also applies to individuals accompanying the armed forces, which may include a dependent of a DOD contractor or subcontractor employee. This law authorizes DOD law enforcement personnel to arrest suspected offenders IAW applicable international agreements and specifies procedures for the removal of accused individuals to the US. It also authorizes pretrial detention and the appointment of counsel for accused individuals. It should be emphasized, however, that the law does not extend UCMJ jurisdiction over contractor employees. Actual prosecutions under the Military Extraterritorial Jurisdiction Act are handled by federal civilian authorities.

4-47. In addition to the above, the military commander can indirectly influence the discipline of contractor employees through revocation or suspension of clearances, restriction from installations or facilities, or revocation of exchange privileges. The process of removing contractor employees from the AO is dependent upon the policies issued by the combatant commander and the extent to which those policies are incorporated in the terms and conditions of the contract.

4-48. When confronted with disciplinary problems involving contractors and contractor employees, commanders should seek the assistance of their legal staff, the contracting officer responsible for the contract, and the contractor's management team. This allows for a thorough review of the situation and a recommendation for a course of action based on the terms and conditions of the contract, applicable international agreements, and host-nation or US law.

STATUS OF CONTRACTOR EMPLOYEES

4-49. Nations and their military forces are required to distinguish between military forces (combatants) and civilians (noncombatants), according to the Geneva convention. Contractors are considered to be civilians authorized to accompany the force in the field and, as such, generally cannot be the object of intentional attack. However, they may lose that protection if used in direct support of military operations. Furthermore, contractors may be at risk of collateral injury when they are in close proximity to a lawful target. Therefore, commanders should consider these factors when determining the nature and extent of contractor use, so as not to put them in a position that jeopardizes their status.

4-50. Contractor status is an important issue for the commander in determining the extent of their use and where within the AO they should be permitted. As the environment becomes more hostile, commanders must address a contractor's ability to perform and determine whether or not to provide armed protection. Contractors similarly become concerned about their ability to perform without risk of physical harm or capture and the extent of their rights if captured.

Hague and Geneva Conventions

4-51. Law of war treaties, such as the Hague and Geneva conventions, attempt to establish and clarify the status of contractors when supporting military operations. These treaties entitle contractors to be treated as prisoners of war.

4-52. The 1949 Geneva convention relative to the treatment of prisoners of war (Article 4) provides that prisoner-of-war status upon capture is extended to, among others, contractor employees.

4-53. During armed conflict with nations that are not signatories to these treaties, the status of contractors may be less clear. Commanders should consult their staff judge advocate for legal advice in these situations.

Hague Convention in 1907 (Article 13)

"Individuals who follow an army without directly belonging to it, such as...contractors, who fall into the enemy's hands and whom the latter thinks fit to detain, are entitled to be treated as prisoners of war, provided they are in possession of a certificate from the military authorities of the army which they were accompanying."

1949 Geneva Convention, treatment of prisoners (Article 4)

"Persons who accompany the armed forces without actually being members thereof, such as ...contractors, who fall into the hands of the enemy, and whom the latter think fit to detain, shall be entitled to be treated as prisoners of war, provided they have received authorization from the armed forces which they accompany, who shall provide them for that purpose with an identity card."

Status-of-Forces Agreements

4-54. Status-of-forces agreements may cover a wide range of issues impacting on the use of contractor support during military operations. Essentially, these agreements establish the legal obligations to be followed when operating within or in-transit through a particular nation. When applicable, these agreements may establish legal obligations independent of contract provisions. Status-of-forces agreements also define the legal status (e.g., host-nation criminal and civil jurisdiction) and legal obligations (e.g., taxes, customs, etc.) of contractors and contractor employees in a host nation. Commanders planning the use of system or external support contractors (i.e., US or TCNs) must consider including them in status-of-forces agreements. In addition to employees status, items to be considered include customs, taxes, and documentation of technical expert status. Status-of-forces negotiations may also involve countries that must be transited to reach the AO.

CONTRACT PROVISIONS

4-55. The SOW prepared by the requiring unit or activity states what support is needed. This document identifies all of the known requirements for the contractor to perform his mission and is the document, along with the applicable OPLAN/OPORD, upon which a contract is based. The contract provisions direct a contractor's work and establish the legal and binding agreement with the government. Contracting professionals, using federal and departmental acquisition regulations, write the contract based on the requiring unit or activity's SOW. Chapter 2 provides a detailed discussion concerning contract provisions and communicating requirements.

Chapter 5

Supporting

INTRODUCTION

5-1. Contractors are expected to be self-sufficient, handling all actions necessary to perform under the terms and conditions of the contract without significant assistance from the government. However, in some circumstances, it is appropriate and necessary for the government to provide support to a contractor when deployed into an AO. Government-furnished support may be appropriate if it is less expensive than contractor-provided support. Additionally, government-furnished support may be necessary when the government controls the support needed or is the only source of support, such as transportation within the AO. Figure 5-1 illustrates the level of support that may be provided to a contractor based on factors, such as nature of the operation and location on the battlefield. Government requirements for support to a contractor must be identified during the planning process and included in the applicable contract. Appendix B provides a notional example of government-support planning requirements.

5-2. Typically, in a contingency or battlefield environment, government-furnished support to a contractor is either organizational or life support. Organizational support includes transportation, facilities, and equipment and materiel necessary for the contractor to perform its mission. Life support includes lodging and subsistence, morale support services, medical care, mortuary affairs, as well as religious, legal, and postal services that either the contractor cannot provide or for which the government is better suited. Force protection is another form of support that the government provides and is addressed in detail in Chapter 6.

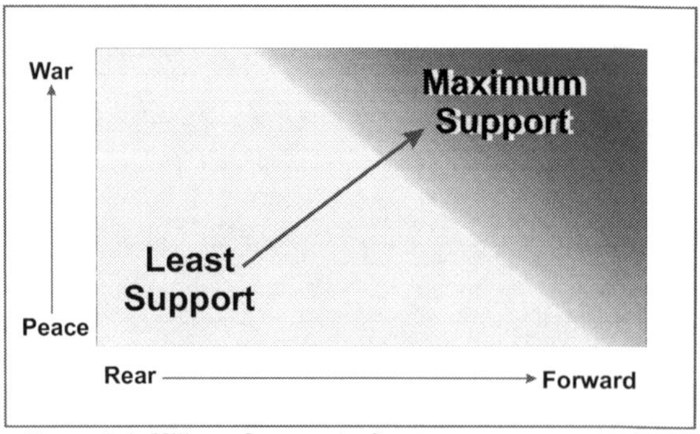

Figure 5-1. Military Support to Contractors

5-3. The nature and level of government-furnished support provided to contractors depends on the point of origin of the contractor. For US contractors and contractors originating from outside the AOR, the level of support will be greater because they may require transportation to get to the theater and facilities to operate once they are in place. On the other hand, contractors residing in the AOR may only require transportation or access to the LOCs, force protection, or no support at all.

ROLES AND RESPONSIBILITIES

5-4. As with the other aspects of contractor-provided support, the responsibility for ensuring that contractors are considered for and receive the requisite government-furnished support involves commanders and staffs at all levels, the contracting community, and the contractor. Government-furnished support to contractors, whether it is organizational or life support, may be key to the contractors' successful performance. Without the support, contractors may find it difficult or impossible to operate in the battlefield environment, and, when a contractor does not perform, the government may place itself in a position of liability, including entitlement to monetary compensation, not to mention the operational impact of the loss of contractor support. Likewise, if contractors are directed to provide their own support, the government can expect higher contract costs as a result.

ASCC/ARFOR

5-5. In keeping with his overall responsibility to prepare, train, equip, administer, and provide Army forces to the combatant commander for an operation, the ASCC commander provides support to the contractors supporting Army operations. Because the ASCC commander, and the subordinate ARFOR commander (if applicable), has the responsibility to arrange and plan CS and CSS, it is at this level that specific units or organizations (possibly including a contractor) are identified and tasked to provide the necessary support to contractors. The ASCC/ARFOR tasks those units that are in the best position or have the best capability to provide the required support. Often the units tasked are those that have a habitual relationship with the contractor or the ones that will be managing the contractor in its execution of the contract.

CONTRACTING ACTIVITIES

5-6. The various contracting activities supporting an operation ensure that the type, manner, and scope of support that is to be provided to a contractor is incorporated into the applicable contract. These activities also advise the commander as to the extent that support to be provided is appropriate and can legitimately be provided. Once included in the contract and communicated to the contractor, these contracting activities ensure that support is being provided either through their own resources or through contract administration organizations, such as DCMA, or support commands such as USAMC.

SUPPORTING ORGANIZATION

5-7. The supporting organization is that organization or activity tasked to furnish organizational or life support to a contractor. It may be the organization that the contractor is providing support to, the requiring unit or activity, a functional organization (such as a transportation unit) or the USACE that would provide equipment or facilities, or any other organization that possesses the support capability. The supporting organization provides specified organizational support or life support, and may, if appropriate, use another contractor to provide such support.

CONTRACTOR

5-8. A contractor's responsibility regarding support provided by the government is to ensure that it complies with the terms and conditions of the contract concerning government-provided support. Contractors are obligated to follow organizational support guidelines provided by the government regarding the accountability and proper and safe operation of vehicles and equipment. They are required to adhere to cleanliness standards when occupying government-furnished lodging or work areas and to comply with established procedures for the use of government-furnished life support. Contractors must follow the same procedures and guidelines as those mandated for military and DACs.

OPERATIONAL SUPPORT TO CONTRACTORS

5-9. Operational support provided to contractors is that support that enables the contractor to perform its support mission. This type support normally includes equipment and services, facilities, transportation and personnel recovery. Operational support is provided by the government when the contractor does not possess the unique military capability, because it is more economical, or to prevent competition for limited resources.

GOVERNMENT-FURNISHED EQUIPMENT AND SERVICES

5-10. The general policy of the government is that contractors furnish all equipment and services required for the performance of a contract. However, the government can provide equipment and services when necessary to achieve significant savings, standardization, expedited performance, or when it is otherwise in its best interests. Potential savings, military-unique equipment or services, and the degree of government control over the equipment or services influence the requirement for the military to provide GFE and services to a contractor. Contracts that provide for GFE to contractors must specify what support, property, and training the Army will furnish to the contractor prior to commencement of an operation. Contracts must also specifically address responsibility for storage, maintenance, accountability, and testing for all furnished vehicles and equipment.

5-11. GFE must be properly controlled. The lending activity designated to furnish the equipment and the supporting contracting element must establish and maintain accountability over the equipment as long as the equipment is required or until the contract expires. Procedures for controlling and accounting for equipment provided to contractors is contained in

AR 710-2 and AR 735-5. For equipment or materiel acquired by the contractor during the performance of the contract, ownership will pass to the government, in accordance with Army and contracting procedures, when the contract is completed or terminated.

5-12. Operational support services provided to contractors may include connectivity to telecommunications resources, fuel for contractor-owned or government-furnished vehicles, or any other service that might be furnished to assist the contractor in performing its mission. Services offered must be planned for in advance so that sufficient capability is available to ensure that support to military forces is not degraded and to allow the contractor sufficient time to arrange its own support if the government is unable to do so. Advance planning also provides information to the contracting officer essential to negotiate the price of the contract.

FACILITIES

5-13. During the planning process, commanders and staff planners must address the need for contractor facilities. Contractors, primarily external support and system contractors, supporting a military force in an AO often require facilities and/or real estate to perform their work. Depending on the operational situation, either the contractor or the government arranges the use of real estate and facilities as required. Determining who is best suited to make the arrangements should be made during the planning process with the following considerations:

- What type of facilities does the contractor require?
- What is the cost if the contractor provides his own facilities?
- Should contractor operations be in a secure location?
- Are there enough facilities to accommodate both military and contractor needs, thus avoiding unnecessary competition?
- Does the military control all facilities in the area?

5-14. Commanders and staff planners should be concerned with cost, physical protection, base access, and coordination of contractor and military requirements. Contractor competition for limited facilities may inadvertently affect their cost. The solution is for the military to either assume control of all facilities and provide what is needed to the contractor or establish coordination procedures in the contract to avoid unnecessary competition for facilities.

5-15. Contractor requirements for facilities, when provided by the government, are passed to the theater real estate manager for action. The theater real estate manager may be the ASCC's operations section terrain manager or a USACE corps real estate support team (CREST). A contractor's specific facility requirements are not normally identified during the planning process, but during the negotiation process prior to contract award. For this reason, operational planners must be consulted during the negotiation process so that contracting professionals do not obligate the government for support that may not be available. When the government does not have the capability to support a contractor's needs, it should be prepared to bear the cost of contractor-obtained facilities.

TRANSPORTATION

5-16. Contractors may have difficulty obtaining transportation when deploying to, and operating in, an AO. In many operations, US forces control transportation assets and port facilities and dominate or control many of the associated services. To ensure contractors are present in the AO at the proper time and have the necessary services to perform their contract, the Army must determine whether it, or the contractor, provides the required transportation. When a contractor does not have full control over the circumstances that affect how and when its transportation support is provided, it may be in the Army's best interest to assist.

5-17. The extent that government-furnished transportation is provided to a contractor is based on the operational environment and the government's capability to meet the contractor's specific needs. These needs not only include the deployment or movement of the contractor to the AO, but also those to satisfy operational needs during the operation itself. Contractor transportation requirements may include the following:

- Air, rail, and sea transport to move personnel and equipment to the operational theater.
- Port handling equipment or services
- Wheeled or aerial transportation support to move supplies, equipment and/or personnel within the AO.
- Government furnished vehicles to transport personnel and equipment in the AO.
- Access to LOCs and movement clearances in the AO.

5-18. When government-furnished vehicles are provided, the necessary maintenance and fuel support should also be included. Likewise, when contractors provide their own vehicles, fuel and maintenance support must be considered and provided when it is in the government's best interest to do so.

5-19. Contractors must be aware that insurance companies may limit or deny coverage on private insurance policies if an individual is being transported on military transportation, such as rotary-wing aircraft. When this occurs, the government will underwrite insurance to allow coverage to continue. Planners at all levels must provide for situations when contractors or the military have to obtain special coverage.

PERSONEL RECOVERY

5-20. It is DOD policy to protect its personnel (to include contractors who deploy with the force), prevent the explotation of its personnel by adversaries, and reduce the potential for captured personnel being used as leverage agains the United States. It is important for commanders and eligible contractor personnel are aware of this policy. For more information see DOD Directive 2310.2, *Personnel Recovery* and DOD Instruction 2310.4, *Reparation of Prisoners of War (POW), Hostages, Peacetime Geovernment Detainees and Other Missing or Isolated Personnel.*

OTHER SUPPORT

5-21. Additional government support to contractors is dependent on the operational environment and the type of contractor involved. For contractor employees deploying with the force, support, such as lodging and medical treatment; subsistence; laundry and shower; medical; mortuary; morale, welfare and recreation (MWR); postal; and religious, mortuary affairs, next of kin notification, is generally similar to that provided to DACs participating in the same operation. However, some support such as legal support is not the sam as for DACs. In some cases, contractors may also live and work under field conditions similar to those for the supported military forces. When contractors are placed in this type environment, furnishing their own basic life support becomes difficult or impossible. As a result, the military must make provisions for their support or must not use contractors. Regardless of the source, contractor or military, life-support requirements must be identified and included in OPLANs/OPORDs and the governing contract. In addition to cost, considerations necessary to determine whether the contractor or the military should provide the support include:

- Ability to provide the support.
- Potential for competition between the military and contractors for support resources.
- Potential for establishing redundant support structures.

LODGING/SUBSISTENCE

5-22. Contractors should provide their own lodging and subsistence unless the operational environment dictates otherwise. The circumstances under which the military provides this support would be those in which the contractor has no commercial infrastructure from which to draw from or when the cost for a contractor to furnish the support is not economical. In those situations when contractor-arranged housing would impede the government's efforts to provide force protection, generate competition with the military, or adversely influence prices, the military must consider providing the support. The ARFOR commander retains the authority to direct where contractor employees reside, within the terms and conditions of the contract, but must utilize the same standard that is applied to DAC personnel of similar grade and responsibility level (e.g., a contractor technical representative or first-line supervisor would be treated the same as a GS 11-13).

5-23. During military operations, contractor employees often live under field conditions. Field conditions are quite different from normal civilian life and are characterized by austere and communal living and a collective responsibility for the living area. Contractor employees may experience a general lack of privacy with lodging normally consisting of large communal tents. While in this field environment, contractor employees will be expected to maintain a clean living area, be considerate of others, and adhere to the same rules, policies, directives, and general orders as the military and DAC personnel.

5-24. Subsistence may be provided to contractors, either in conjunction with government-provided lodging, or separately, when contractor employees during their daily work shift are unable to obtain subsistence. The food provided might be pre-packaged rations with very little opportunity for choice; consequently, special diets may not be accommodated.

5-25. Although it is natural to expect reimbursement from contractors for the cost of lodging and subsistence, the cost for such support would normally be included in the overall cost of the contract. Therefore, providing the support should be done on a nonreimbursable basis, eliminating the unnecessary administrative burden of reimbursement. However, military planners must include the cost of supporting contractors in the overall cost of the operation so that adequate funding is provided.

POSTAL

5-26. The nationality of the contractor employee usually determines postal support. US citizen contractor employees, who deploy in support of US armed forces, may be authorized use of the military postal service (MPS) if there is no US postal service available and if MPS use is not precluded by the terms of any international or host-nation agreement. Local nationals hired in-country by DOD, or subcontracted by a DOD contractor, or TCN employees normally are provided with postal support through the existing host-nation system or through arrangements made by the employing contractor.

5-27. The participating ASCC/ARFOR personnel planners are the best source for advice concerning available postal support. From an analysis of the AO, force composition, and international and host-nation mail capabilities, they can offer an informed recommendation on the level of support to be provided. The effectiveness of postal support is directly related to the personnel accountability system in place in the AO. The contractor accountability system being used must interface with the AO postal system.

MORALE, WELFARE, RECREATION

5-28. Maintaining an acceptable quality of life is important to the overall morale of any organization, including contractors. Deployed contractors have a responsibility to provide MWR and other quality-of-life support to their employees as much as practical. Theater support contractors generally are not provided such.

5-29. The military may provide MWR support to contractor employees when contractor sources are not available, subject to the JTF/ARFOR commander's discretion and the terms of the contract. The availability of MWR programs in an AO vary with the deployment location. MWR activities available may include self-directed recreation (e.g., issue of sports equipment), entertainment in coordination with the United Services Organization and the Armed Forces Professional Entertainment Office, military clubs, unit lounges, and some types of rest centers.

5-30. US citizen contractor employees may be eligible to use Army and Air Force Exchange Service, Navy Exchange, or Marine Corps Exchange facilities for health and comfort items. Use of these facilities is based on the combatant commander's discretion, the terms and conditions of the contract, and any applicable status of forces agreements.

5-31. Regardless of who provides MWR support, governing OPLANs/OPORDs and contracts must identify contractor support needs and designate responsibility. These documents should also articulate the extent MWR support is authorized so that employing contractors can arrange for services that will not be provided.

HEALTH SERVICES

5-32. Medical and dental care is provided to contractors as specified in existing DOD and Army policy and as outlined in the contract. Eligible contractor employees receive support equivalent to that provided to military personnel. This care includes—

- Inpatient and outpatient services.
- Evacuation.
- Pharmaceutical support.
- Any other medical support as determined by appropriate military authorities, in accordance with recommendations from the command surgeon.

5-33. Eligibility for this care normally is extended to US citizen and TCN system and external support contractor employees. Local national employees receive emergency care only when injured on the job and working within military unit areas. For example, a baker injured while working in his own shop would not be eligible; a truck driver, operating his own vehicle, and injured while on a supply distribution mission transporting military supplies would be eligible. If emergency care is provided to local national contractor employees, they will be transferred to a local medical facility as soon as it is medically feasible. Routine medical and dental care are not provided to theater support contractors or sub-contracted local-hire external support contractor employees. These individuals must obtain routine care from the local infrastructure. Army medical personnel provide all occupational health and preventive medicine inspections, screenings, and remedial treatment to contractor employees, to include animals as appropriate, regardless of their status.

5-34. As forward-deployed field medical organizations are neither organized nor equipped to process billings or receive payments for medical services, reimbursement from eligible contractor employees or third parties is not practicable or cost effective. However, where a contractor employee is medically evacuated from the AO to a medical treatment facility (MTF) funded with defense health program dollars, normal reimbursement policies apply for medical services provided by that MTF. This includes instances where an MTF is outside the AO but still in the same theater.

5-35. The Army medical department provides pharmaceutical support to deploying employees of system and external support contractors. As part of the predeployment process, the pharmaceutical requirements of each contractor employee is determined. Medications listed on the basic core formulary (BCF) are continued and refilled in theater from the appropriate Army medical organization.

5-36. All contractor employees are required to bring a 90-day supply of personal medications into the AO. If a refill of a medication is required, authorized contract employees contact the nearest medical unit with pharmacy services. Refills are normally processed when there is a 15-day supply or less on the prescription.

5-37. Medical personnel are responsible to evaluate contractor employees with prescribed medications that not listed on the BCF to determine if a BCF medication can be substituted. When this is possible, a written prescription for the BCF medication is given to the contractor employee, and they are advised to obtain a 90-day supply at their own expense. If there is no clinically acceptable BCF substition available, the contractor will be required to make arrangements to obtain this medication in theater from civilian sources or to be devilered from CONUS. If this is not feasible, then the contractor employee in question should not be allowed to deploy.

5-38. If eyeglasses are required, contractor employees will deploy with two pairs of glasses and a current prescription. The military provides eyeglass inserts for use in chemical protective masks as part of the predeployment process. Emergency replacement of glasses is the responsibility of the military medical logistics system. Routine replacement is the responsibility of the employee/employer.

5-39. Routine and emergency medical support to eligible contractor employees is provided on an area basis from appropriate Army medical organizations. System contractors operating within a division AO receive medical support from appropriate divisional medical elements.

5-40. Intratheater evacuation of US citizen and TCN contractor employees is accomplished using military medical evacuation air and ground vehicles, unless provisions for alternate evacuation means are provided or planned. Medical evacuation out of the AO is the responsibility of USTRANSCOM. This movement is coordinated between the theater patient movement requirements center, the global patient movement requirements center, and the TACC. These organizations determine the best lift available to move the patient out of the theater.

5-41. The remains of system- and external support contractor employees who die while patients in a military medical treatment facility are handled by appropriate mortuary affairs organizations. Mortuary affairs is discussed later in this chapter.

5-42. Limited support to security/working dogs under contract includes limited emergency care, preventive medicine inspections, and necessary prophylaxis. The contractor ensures that all animals meet the standards of the host country, as well as the standards enforced by the US military for working animals. The contractor ensures all animals deploy with necessary routine medications and chemoprophylaxis. The contractor also ensures that these medications and chemoprophylaxis are resupplied at the contractor's expense.

5-43. Upon redeployment, contractor employees receive medical and dental screenings to identify and document any medical problems that may be connected with the deployment. These redeployment screenings may be completed prior to AO departure, at a CONUS IDS/CRC site, or at the home of record.

RELIGION

5-44. When contractors are deployed in support of military operations they are provided religious support in accordance with OPLAN/OPORD religious support annexes and appendices. The appropriate religious support section assesses the requirements and includes them in the religious support plan. US contractor family members may seek and expect to be provided religious support through CONUS installations while spouses are deployed in support of military operations (garrison commander approval may be required for access when military installations are closed to civilians). Unit ministry teams will make every effort to plan for adequate religious support resources, including literature and sacramental supplies, to support contractor employees who deploy with the force.

MORTUARY AFFAIRS

5-45. The joint mortuary affairs program is a broadly based military program that provides for the necessary care and disposition of deceased personnel, including personal effects, across the full range of military operations. Combatant commanders control and coordinate mortuary affairs operations for the search, recovery, tentative identification, care, and evacuation or temporary interment, disinterment, and reinterment of deceased personnel within their AOR.

5-46. Under the joint mortuary affairs program, all contractor employees who are in direct support of military operations and who die in the line of duty are eligible to receive mortuary affairs support on a reimbursable basis. The specific nature and extent of the support is determined during the planning process and communicated to military forces and contractors through governing OPLANs/OPORDs and contractual documents. The responsible support commands and related ASCC/ARFOR staff maintain visibility over contractors work in coordination with mortuary affairs and personnel units when contractors require support.

NEXT OF KIN NOTIFICATION

5-47. In the event a contractor employee dies or is missing, next of kin notification is made by the:

- Military when the contractor employee involved is a US citizen. The military uses information provided by the contractor during predeployment or reception processing. In some cases, the military may affect notification through the employing contractor.
- Employer when the contractor employee is a citizen of the host nation or a TCN.

LEGAL

5-48. Army policy is that contractor employees in the US preparing to deploy, or those already deployed, generally are not eligible to receive legal assistance from Army military or civilian lawyers. Legal assistance is the responsibility of the contractor. The contractor ensures that its employees satisfy all legal requirements, such as last will and testament, guardianship arrangements for children, estate planning, and any other needs prior to departure.

5-49. There is, however, an exception to this policy. Contractor employees may receive certain legal assistance when the government is contractually obligated to provide this assistance as part of their life support. When this is the case and the legal assistance is provided overseas, it must be in accordance with applicable international agreements and approved by the host-nation government. Legal assistance, when provided, is limited to ministerial service (e.g., notarial services), legal counseling (including the review and discussion of legal correspondence and documents), legal document preparation (limited to power of attorney and advanced medical directives), and assistance in retaining civilian legal counsel.

Chapter 6

Force Protection

INTRODUCTION

6-1. JP 3-0 and FM 3-0 define force protection as "actions taken to prevent or mitigate hostile actions against DOD personnel, resources, facilities and critical information. These actions conserve the force's fighting potential so it can be applied at the decisive time and place, and incorporates the coordinated and synchronized offensive and defensive measures to enable the effective employment of the joint force while degrading opportunities for the enemy. Force protection does not include actions to defeat the enemy or protect against accidents, weather, or disease." Reducing the probability and minimizing the effects of enemy action on personnel, equipment, and critical facilities can accomplish force protection.

6-2. Force protection is a deliberate and dynamic process that begins long before the actual conflict. Its character may change in different situations, but the process remains the same: an understanding of threat and the development of a system of indications and warnings that will facilitate a proactive, predictive response to enemy and terrorist action. Force protection may include fortification construction, electronic countermeasures, integrated air defense coverage, NBC defensive measures, and rear operations to include specific antiterrorist actions. Force protection and antiterrorism measures are meant to confuse and deceive the enemy so that military forces and accompanying civilian personnel are less vulnerable from attack. Force protection responds quickly to enemy attacks with protective measures that enhance our military force survivability and ensure the rapid recovery and restoration of combat power.

6-3. Force protection and antiterrorism actions include not only military and DAC personnel but contractor employees as well. Contractor employees, because of their status as civilians authorized to accompany the force in the field, bring with them an inherent need for force protection. Contractor employees cannot be required to perform force protection functions described above and cannot take an active role in hostilities but retain the inherent right to self-defense. Because of their civilian background, they may possess neither the training nor experience to actively participate in force protection measures, and the rules governing warfare preclude them from doing so except in self-defense. Therefore, the Army's policy has become that when contractors are deployed in support of Army operations/weapon systems, they will be provided force protection commensurate with that provided to DAC personnel. Commanders must understand that contractors are subject to the same threat as soldiers and must plan accordingly. Contractors, when placed in a position of risk, must be protected, or the support they provide may be degraded. Contractor-related force protection measures must be addressed in the planning process. Appendix B provides a notional example of contractor force protection planning requirements.

ROLES AND RESPONSIBILITIES

6-4. Protecting contractors and their employees on the battlefield is the commander's responsibility. When contractors perform in potentially hostile or hazardous areas, the supported military forces must assure the protection of their operations and employees. The responsibility for assuring that contractors receive adequate force protection starts with the combatant commander, extends downward, and includes the contractor.

6-5. The government is responsible for providing a safe workplace that enables the contractor to perform unhindered by circumstances beyond his control. Commanders and staff planners must assess the need for providing force protection to a contractor and designate forces to provide appropriate security. The mission, threat, and location of contractor operations determine the degree of force protection needed.

6-6. Protection for contractors involves active use of armed military forces to provide escort or perimeter security, and passive measures that include protective military equipment, training, and equipping of contractor employees in self-protection (NBC and weapons).

GEOGRAPHICAL COMBATANT/SUBORDINATE UNIFIED COMMANDER

6-7. The combatant commander makes the decision to provide force protection to participating contractors. These commanders and staffs, through the risk assessment process, evaluate the force protection implications of contractor involvement according to the nature and extent of the operation. This risk assessment, based on METT-TC, considers such factors as contractor location on the battlefield, availability of security forces, enemy capabilities, and the criticality of the contractor's functions. (For a detailed discussion of risk management, refer to FM 3-100.14; see FM 5-0 for information on risk assessment in the planning process.) Also, the OPLAN/OPORD and supporting contracts must identify the specific force-protection requirements and designate the forces to handle them. This allows units and contractors to properly integrate support requirements into their operational procedures.

ASCC/ARFOR

6-8. One of the ASCC's most important responsibilities is to conserve the fighting potential of his force so that it can be applied at the decisive time and place. Force protection includes protecting the force, including contractors, from enemy air, ground, and sea attack. The combatant commander may delegate to the ASCC (or ARFOR when the ASCC is not deployed) as the joint rear area coordinator (JRAC) responsibility for the rear area where the majority of contractors would be located. In this capacity, the ASCC, normally through its ASCC rear headquarters, coordinates between all Services to facilitate a secure environment to allow for uninterrupted sustainment, host-nation support, infrastructure development, and force movements. See JP 3-10 and JP 3-10.1 for further information on JRAC functions.

CONTRACTING ACTIVITIES

6-9. Responsibility for ensuring that the combatant commander's force protection decisions are included in the appropriate contracts and communicated to the contractor lies with supporting contracting activities. These contracting activities ensure, either through their own resources or through contract administration organizations (e.g., DCMA), that support is provided and the contractor is complying with established force-protection policies and procedures.

SUPPORTING ORGANIZATION

6-10. Commanders at all levels must designate specific supporting organizations responsible for providing force-protection support to contractors. To do this, they must have the requisite visibility over the contractor and contractor employees supporting the force. Normally, commanders designate the supported organization or other unit to provide this protection.

CONTRACTOR

6-11. Contractors ensure that all of their employees follow all force-protection requirements and supporting organization policies stated in the contract. Contractors (when required and authorized) should, as a minimum, ensure that their employees receive the directed NBC protection and weapons familiarization training. Additionally, contractors should inform them of local and national threats, facility and travel security requirements, and off-limits restrictions directed by the commander. Contractors are expected to take passive force-protection measures for the safety and security of their employees. Also, contractors should mandate measures for self-defense such as conducting driving classes, issuing cell phones, and establishing procedures for reporting suspicious incidents.

FORCE PROTECTION GUIDELINES

6-12. The degree of contractor force protection depends on the situation; key factors are the nature of the operation, location on the battlefield, and the level of threat. The following force-protection guidelines, based on the DOD terrorist threat levels found in DOD O-2000.12-H (draft), recommend the minimum protection and security required for contractors and are categorized by the level of threat.

NEGLIGIBLE THREAT

6-13. A negligible threat is one where a capability to conduct an attack may or may not be present.

6-14. This threat level is normally associated with such peacetime operations as disaster relief and most humanitarian-support operations. At this threat level, risks are the same as those associated with normal, everyday operations. With no enemy action expected, there would be no change to the normal operational routine of the contractor. Generally, only normal safety practices would be required. Contactor employees would face no restrictions in operational location, movement, etc.

LOW THREAT

6-15. A low threat is one where a capability is present but demonstrated activity may or may not be present.

6-16. This threat level is associated with some humanitarian support and peacekeeping operations. A low threat level is greater than the threat in normal peacetime operations and involves external dangers. Minimum measures would require a military escort for contractor employees working away from the main operating base or group.

MEDIUM THREAT

6-17. A medium threat is one where a capability of demonstrated activity is present, but intent to conduct threat activities may or may not be present.

6-18. This threat level expects casualties and limited indiscriminate combat action not specifically aimed at the military force. Local violence and terrorist action may exist in a generally unstable area. In addition to those implemented for a low threat, force-protection measures include armed military-escort convoy protection for groups working away from a protected area. At this juncture, leaders should start evaluating the need to evacuate non-essential civilians and relocate to a more secure area. Contractor operations should be reduced to the minimum required to maintain existing levels of service without incurring degradation. Planned and routine maintenance should be assumed by military personnel or delayed, if possible. Where delay of repairs or maintenance may impair fighting or defense capability, contractors should remain. Military personnel should replenish or assist in replenishing expended materiel relating directly to the mission. Routine resupply should be delayed or consolidated.

HIGH THREAT

6-19. A high threat is one where a capability of demonstrated activity history and intentions is present.

6-20. This threat level is associated with peace enforcement and major theater war operations where open combat may occur. Casualties are likely at a high-threat level, but it is short of open warfare. Action is directed at participating military forces, with associated casualties being a result of both direct and indirect action. The operational situation is unstable and may deteriorate. Minimum measures, above those implemented for a medium threat, require issuing personal protective equipment and transporting personnel in protective vehicles. If possible, civilians are kept in secure and protected areas; only essential contractors would remain in place. Contractors are permitted to undertake repairs to mission-critical systems and equipment, the loss of which would degrade the fighting/defensive capability of the unit. Additionally, logistical support staff needed to maintain the flow of sustainment materiel without which the mission would fail may remain in place. If practical, military personnel assist in some of their tasks. Where feasible, military personnel undertake maintenance responsibilities if contractor employees are removed from the theater.

CRITICAL THREAT

6-21. A critical threat is one where a capability of demonstrated activity is present, and current credible activity information indicates preparation for a specific anti-US operation, but historical or recent demonstrated anti-US activity may or may not be present.

6-22. Critical threat describes open warfare with casualties inflicted by opposing forces and dangers introduced externally. Minimum measures would require the same as those implemented for a high threat, with movement restricted and within armed formations. Secure protective shelter should be used for all support operations. At this level, contractors remain in the AO only if it is an absoulute necessity, when contractor-provided support cannot be replaced with military capabilities. Note also that contractors cannot be forced to remain in theater against their will, as they are not subject to the UCMJ (except in formally declared war). Refusal to perform, however, may result in civil lawsuit for breach of contract.

ACTIVE FORCE-PROTECTION MEASURES

6-23. Active force-protection measures involve the overt use of military forces to assure contractor operations are protected from enemy attack. These measures involve the dedicated use of military forces to provide perimeter security for contractor work areas and armed escort for contractor movements around the battlefield. Active force-protection measures are employed when contractors are operating in locations where a medium to critical threat exists, and are provided at the specific direction of the combatant commander because of the critical nature of the support the contractor is providing.

6-24. The scope and nature of active force-protection measures are addressed during the risk assessment portion of the planning process. When the need for contractor support exceeds the risks associated with their use, active force-protection measures are implemented. These measures include locating contractor-support operations in secure areas or facilities, where disruption to their activities is minimized. When contractors are collocated with the military forces they support, they are included in the units overall force-protection plan. Otherwise, contractors may require dedicated protection in the form of military police or infantry forces.

6-25. Contractor movements must also be protected when the nature of the support being provided is critical to the success of the operation. Contractors should either be provided with armed escorts, incorporated into military convoys traveling to the same destination, or rely on military forces to move their equipment, personnel, and supplies. When considering the best option for contractor movement, the government must be sensitive to the possibility of relieving the contractor of responsibility for providing support when it employs force-protection measures that may restrict the contractor's ability to perform.

PASSIVE FORCE-PROTECTION MEASURES

6-26. In addition to active force-protection measures that may be implemented, a number of passive measures are available to contractor employees and supported units.

MILITARY CLOTHING AND EQUIPMENT

6-27. Contractor employees supporting military operations should be visibly distinct from the forces they are supporting so that they do not jeopardize their status as civilians authorized to accompany the force in the field. Unless specifically authorized by the combatant commander, contractor employees will not wear military uniforms or clothing except for specific items required for safety or security, such as chemical defense equipment (CDE), cold weather equipment, or mission-specific safety equipment. Either the government or the contractor may decide that a uniform appearance is necessary for contractor employees. In this case, the contractor provides appropriate attire which is distinctly not military and which sets them apart from the forces they are supporting.

CHEMICAL DEFENSE EQUIPMENT

6-28. The combatant commander determines the contractor's requirement for CDE for NBC defense, immunizations (such as anthrax vaccine), and training, which is theater-specific and dependent upon the threat and nature of the duties. The environment and potential use of a nuclear, chemical, and/or biological warfare weapon is the basis for determining the need to equip and train contractor employees with NBC equipment and survival skills. The combatant commander's IPB provides the analysis for potential use of NBC weapons. Once completed, the combatant commander issues guidance for NBC protective postures. When approved for contractor employees, training and equipment are dependent upon the threat and nature of the duties. Either the deploying force-projection platform, CRC/IDS, or an in-theater processing center provides individual protection equipment and training when planners determine it to be in the best interest of the government and incorporate it into the terms of the contract.

WEAPONS

6-29. The decision to allow contractor employees to carry and use weapons for personal protection rests with the combatant commander. The general policy of the Army is that contractor employees will not be armed. However, under certain conditions dictated by METT-TC, they may be allowed to arm for self-defense purposes. Once the combatant commander has approved their issue and use, the contractor's company policy must permit its employees to use weapons, and the employee must agree to carry a weapon. When all of these conditions have been met, contractor employees may only be issued military-specification sidearms, loaded with military-specification ammunition. Additionally, contractor employees must be specifically trained and familiarized with the weapon and trained in the use of deadly force in order to protect themselves. Contractor employees will not possess privately owned weapons. When determining to issue weapons to a contractor the combatant commander must consider the impact this may have on their status as civilians authorized to accompany the force.

TRAINING

6-30. Contractor employees deploying either from CONUS or OCONUS may require specific force-protection training to prepare for the environment in the AO. Training requirements must be identified during operational planning and be included in the contract. The type of contractor (theater support, external support, or system) influences the nature and degree of the training to be provided. Training may include—

- Geneva convention.
- Code of conduct.
- Health and sanitation.
- Applicable status of forces agreements.
- Security.
- Weapons familiarization and use of deadly force.
- NBC training, including protection from toxic industrial materials.
- Survival, escape, resistance and evasion (SERE)
- Rules of engagement.
- Host-nation laws.

6-31. The contractor or the Army, whichever is specified in the contract, can conduct the required training. In areas such as weapons familiarization or chemical defense, the Army may be the more appropriate source. As such, the deploying force-projection platform, CRC/IDS, or the in-theater reception point would conduct the training in conjunction with other processing activities.

FORCE PROTECTION/ANTITERRORIST THREAT FROM CONTRACTORS

6-32. Due to recent terrorist activity against US forces, all units must ensure force-protection/antiterrorism plans and actions are integrated into movement and support operations in all areas. Use of local or TCN contractor employees must be carefully considered from the antiterrorism perspective.

> On 12 October 2000, the USS Cole (DDG 67) suffered a serious terrorist bomb attack in Aden, Yemen, that severely damaged the ship, killing 17 sailors and injuring 39. This attack came in the form of a small boat laden with explosives that, according to some reports, was thought by the crew to have been a part of the scheduled contractor provided refueling support. This attack occurred despite what naval investigators determined were prudent command-directed force-protection measures.

6-33. Commanders at all levels must include the following areas in force-protection/antiterrorism pre-deployment planning:

- Threat and vulnerability assessments. Units assess the threat and their own vulnerability prior to deployment. Assessment must include risk of using local national and TCN contractor employees vice using military or US national civilian capabilities.
- Security planning. Units must take the results of threat and vulnerability assessments and develop security plans when using local and TCN contractors. Tailored intelligence and counterintelligence support, host-nation assistance, and detailed contractor-employee screening plans must be in place. Special empahsis must be placed on local national and TCN contractor-employee access to vulnerable facilities and areas.
- Combat and Combat Service Support Operations. Predictable unit movements and support operations can lead to increased vulnerability of both personnel (to include contracted support) and facilities. Unit commanders must understand that predictability places a higher demand on the unit's ability to know the local threat, assess unit vulnerabilities, and develop self-protection measures to include force-protection/antiterrorism actions as they relate to the use of local national and TCN contractor support.

Appendix A

CONTRACTING OFFICER REPRESENTATIVE GUIDELINES

A-1. A COR is the contracting officer's designated representative who assists in the technical monitoring and administration of a contract. The COR is the requiring unit or activity or supported unit link to the contractor, using the contract administration/management process. This individual is designated in writing and must be a qualified military member (in the grade of E-5 or above)/government employee (according to the requirements established in the Defense Federal Acquisition Regulation Supplement, Part 201.602) to perform the duties and responsibilities delegated by the contracting officer.

A-2. The specific duties and responsibilities of a COR are delegated in writing by the contracting officer. Typically, a COR is responsible for assisting the contracting officer in the following areas:

- Maintaining liaison and direct communications with both the contractor and the contracting officer.
- Monitoring the contractor's performance, notifying the contracting officer of deficiencies noticed during surveillance, and recommending appropriate corrective action.
- Verifying that the contractor performed the technical and management requirements of the contract.
- Performing all necessary inspections.
- Verifying that the contractor corrected all correctable deficiencies.
- Accepting government supplies and services.
- Verifying invoices.
- When applicable, coordinating the deployment/redeployment preparation contractor employees who deploy with the force.
- Assisting in contractor-employee day-to-day management to include visibility and accountability reporting as well as other contractor operations management issues that may affect contractor or unit requirements.

A-3. Although the COR provides a vital link between the military and the contractor, there are certain limits to his authority. A COR is prohibited from:

- Making any agreement with the contractor requiring the obligation of public funds.
- Making any commitments or changes that affect price, quality, quantity, delivery, or other terms and conditions of the contract.
- Encouraging the contractor by words, actions, or a failure to act to undertake new work or an extension of existing work beyond the contract period.

- Authorizing a contractor to obtain property for use under a contract.
- Interfering with the contractor's management prerogative by "supervising" contractor employees or otherwise directing their work efforts.

Appendix B

CONTRACTOR INTEGRATION PLAN ANNEX

B-1. This appendix provides a sample of an OPLAN/OPORD annex oriented on contractors supporting military operations. It is intended to serve as a notional example annex that communicates the commander's directives and guidance concerning contractors supporting an operation. The contractor operations support plan, while not required by doctrine, can serve a valuable purpose. For operational planners and contracting professionals, it can be the single source for commander directives, guidance, and policies regarding how contractors supporting an operation are managed, deployed, supported, and protected. It encompasses all types of contractor support, originating from anywhere in the world, and serves as the critical link between the supported senior commander, the various functional support elements, contracting activities, and the supporting contractors.

NOTE: This sample annex should not be viewed as a source for communicating the Army's overall policy and doctrine for integrating and managing contractors on the battlefield; that is the purpose of AR 715-9 and this manual. To understand the contents of the annex, the reader should read and become familiar with this manual.

B-2. Notional contractor integration plan annex:

Copy_____of_____Copies

Third (US) Army
Fort Gillem, GA
DTG

Annex__ (Contractor Integration Plan) to Third (US) Army Operations Order 01-7527

References:

a. Map Series _____; Sheet Numbers ____.

b. US Central Command Operations Order 02-7527

c. Third (US) Army Field SOP

Time Zone Used Throughout the Order: Local

Task Organization: Annex A (Task Organization)

1. **SITUATION.** See Basic OPORD.

2. **MISSION.** See Basic OPORD.

3. **EXECUTION.**

Third (US) Army Commander's Intent. See Basic OPORD.

a. General. The following are general directives and policies regarding contractor support to this operation:

(1) The contents of this annex will be incorporated into all contracts providing support for this operation.

(2) All units with currently existing contractor support will provide this annex to their supporting contracting activity(ies) for inclusion in their contracts.

(3) All contractors (theater support, external support, and system), supporting this operation and located in the AO, are under the control of and will follow the policies and orders of the Commander, Third (US) Army.

(4) The G-3, in conjunction with primary and functional staff assistance, will provide detailed planning for the management, accountability, and visibility for all contractorssupporting this operation, regardless of origin of the governing contract.

(5) Future contracts awarded in support of this operation will be IAW Appendix __ (Contracting Support) of Annex D (Service Support) and approved by the ARB before any contract award is awarded.

(6) The responsible senior support commander utilizing contractor support will have a contingency plan for all contracts that provide essential services in the event that the contractor is unable or refuses to provide the contracted support. See Appendix __ (Contracting Support) of Annex D (Service Support) for more information.

(7) Government-furnished training is provided in accordance with paragraph 3b (Deployment/Redeployment) of this annex.

(8) Contractors are managed IAW all applicable DOD and Army regulations and paragraph 3c (Management) of this annex.

(9) Government-furnished support is provided in accordance with paragraph 3c (Government-Furnished Support) of this annex.

(10) Contractor force protection is executed in accordance with paragraph 3e (Force Protection) of this annex.

b. Deployment/Redeployment (with appropriate lead and supporting staff).

(1) (G-3) All external support and system contractors, with the exception of the LOGCAP contractor, will deploy under government control. The LOGCAP contractor is authorized to self-deploy, but must be integrated into the TPFDD and obtain arrival slots for its aircraft and vessels.

(2) (G-3) The requiring unit or activity inputs contractor deployment requirements into the TPFDD in accordance with Appendix B, FM 3-100.21. Contractors that have a direct support (habitual) relationship with a specific unit are included in the supported unit's TPFDD.

(3) (G-1) Individual contractor employees deploying from CONUS, with the exception of the LOGCAP contractor, receive predeployment processing and specified training at

the CONUS Replacement Center, Fort Benning, GA. A letter of authorization/identification, issued by the responsible contracting officer, is required. Individual LOGCAP contractor employees receive predeployment processing and training from the LOGCAP contractor IAW the requirements specified in this annex.

(4) (G-1) All contractor employees deploying from OCONUS, with the exception of the LOGCAP contractor, receive predeployment processing and specified training at the theater reception processing center at the APOD/ISB, in Cairo, Egypt. LOGCAP contractor employees deploying from OCONUS receive predeployment processing and training from the LOGCAP contractor IAW the requirements specified in this annex.

(5) (G-1 and Surgeon) All external support and system contractor employees receive HIV, medical, dental, and tuberculin skin test screening during predeployment processing at the designated predeployment processing site. The employing contractor ensures that its employees are in compliance with the health, dental, and physical standards specified in FM 3-100.21 and that they have a doctor's certification that they are fit to perform their duties in the AO. Contractor employees will present their medical/dental records during predeployment processing. Any contractor employees with disqualifying conditions will not be permitted in the AO and are not entitled to any government-furnished support. Theater support contractor employees are excluded from this requirement; they are not authorized government-furnished medical support except that specified in paragraph 3e (Government-Furnished Support) of this annex.

(6) (Surgeon) All external and system contractor employees receive immunizations as specified by the Center for Disease Control (CDC) for the AO. The specific list may be obtained by accessing the CDC Internet-based web site at www.cdc.gov/travel. Employing contractors ensure that their employees receive the following immunizations from their personal physician prior to arriving at the designated predeployment processing site: tetanus, typhoid, yellow fever, polio, MMR, and meningiococcus.

(7) (Surgeon) Contractor employees who are authorized health care support are required to bring a 90-day supply of personal medications. Refills for medication will be furnished by the in-theater medical organization.

(8) (Surgeon) Contractor employees requiring eyeglasses will deploy with two pairs of glasses and a current prescription. Government-furnished eyeglass inserts used in chemical protective masks are provided. Emergency replacement of glasses is furnished by the government; routine replacement is the responsibility of the employing contractor.

(9) (G-3) All external support and system contractor employees, with the exception of the LOGCAP contractor, receive the following training at the designated predeployment processing site. LOGCAP contractor employees receive the designated training from the LOGCAP contractor IAW the requirements specified in this annex.

(a) Geneva conventions.
(b) Code of conduct.
(c) Health and sanitation.
(d) Customs and courtesies for the area of operation.

(e) Status of forces agreements.
(f) Security and force protection.
(g) Operational decurity.
(h) Chemical, biological, and nuclear protection.
(i) Driver and specialized equipment training.

(10)(G-4) All contractor employees supporting this operation are authorized the use of OCIE and CDE in accordance with their location in the AO, the nature of their duties, and the degree of risk involved. The requiring unit or activity identifies specific requirements and obtains the necessary equipment.

(11)(G-1 lead; G-3, G-4, and Surgeon in support) The supporting predeployment processing site accomplishes the following in support of deploying external support and system contractor employees:

(a) Screens and verifies medical requirements, including medical and dental screening, immunizations, DNA screening, HIV testing, and tuberculin skin test screening.
(b) Conducts specified training and briefings.
(c) Issues OCIE and CDE.
(d) Arranges for transportation to the AO IAW TPFDD and USTRANSCOM instructions.
(e) Issues DD Form 1173 (Uniformed Service Identification and Privilege Card), DD Form 489 (Geneva Conventions Identity Card for Persons who Accompany the Armed Forces), personal identification tags, and medical personal identification carrier.
(f) Screens personnel records and enters into the designated personnel accountability system.

(12) (G-1) All deploying external support and system contractor personnel, including those from the LOGCAP contractor, process through the in-theater reception center at the APOD/ISB, in Cairo, Egypt, for verification of accountability and issuing of operational identification. Contractors deploying with their supported units receive reception processing with the unit.

(13) (G-1 lead; G-2, G-3, G-4, Surgeon, and PARC in support) Theater support contractor employees who provide direct support to Army forces and who require access to military facilities are either processed through the Cairo reception center or the designated forward reception center, whichever is closest to their point of origin. The designated reception center accomplishes the following in support of theater support contractor employees:

(a) Conducts limited background/security check.
(b) Conducts specified training and briefings.
(c) Issues OCIE and CDE.
(d) Issues, as required, locally produced ID/access cards.

(14) (G-4) Deploying contractor equipment is processed at the APOD/SPOD IAW TPFDD and USTRANSCOM priorities.

(15) (G-1 lead; G-3, G-4, Surgeon, PARC in support). Redeployment of all contractor employees and equipment, either during the operation or at its termination, is IAW with this annex and instructions issued separately through FRAGO or other means.

(a) All external support and system contractor employees, including the LOGCAP contractor employees, process through the processing center at the APOD/ISB in Cairo, Egypt, prior to departure for accountability and transportation purposes. Personnel who received their predeployment processing at Cairo also receive medical screening and return all government-issued OCIE and CDE. Personnel who received their predeployment processing at the CONUS center receive medical screening and return all government-issued OCIE and CDE there. Theater support contractors process through the same center at which they received their reception processing.

(b) Contractor equipment will be cleaned, processed, and shipped to its destination IAW USTRANSCOM priorities and instructions.

(c) Once redeployment processing is completed, the employing contractor arranges follow-on transportation for its employees.

c. Contractor Management (with appropriate staff lead).

(1) (JAG) Management of contractor compliance with the specific terms and conditions of the contract is the responsibility of the established contracting structure. Commanders are NOT authorized to direct any contractor, or change or modify the terms and conditions of a contract, except through the established contracting system.

(2) (JAG) Management of contractor employees is the responsibility of the employing contractor and does not extend to the military chain of command. Commanders are NOT authorized to direct individual contractor employees except through the contractor's management structure.

(3) (JAG) All requiring activities have a trained COR, appointed by the appropriate contracting officer, to assist in the management of contractor support.

(4) (G-1) Accountability of all contractor employees deploying with the force is through the TPS, the same system used for military and DAC personnel. Specific data requirements are IAW the following:

- Name.
- Sex.
- Social security number (if applicable).
- Date of birth.
- Employment classification (DOD, non-DOD, DOD contractor, nonappropriated funds, Red Cross, other).
- UIC affiliation.
- Marital status.
- Race/ethnic group.
- Religion.
- Passport number (if applicable).
- Point of contact name/phone number.
- Deployment date.
- Blood type.

- Home address.
- Contract number.
- Contractor name.

(5) (All staffs as applicable) All requiring units maintain visibility for contractors providing support to Army forces in the AOR IAW ASCC OPLAN/OPORD guidance. This visibility data is forwarded through the normal chain of command to the ARFOR staff responsible for oversight of that particular contractor function. The ARFOR staff provides contractor visibility information to the ASCC commander/staff as required. Contractor visibility should contain the following information:

- Contract number.
- Contractor name.
- Type of support provided (e.g., equipment maintenance, transportation, construction, food service).
- Name of contractor's in-theater manager and office location.
- Awarding contracting office.
- Contracting officer contact information.
- COR contact information.
- Supported unit (UIC).
- Contractor workforce composition, including: total number of US citizen employees, total number of TCN employees (specify nationality), and total number of local national employees.
- Employee standard name line.
- Employee social security number (or other identification number, such as passport number, for third-country and local national employees).
- Employee home address.
- Employee point of departure.
- Employee AO location.

(7) (G-3) All contractor employees will comply with all command directives, including General Orders #1, covering conduct of personnel.

(8) (JAG) The provisions of the UCMJ DO NOT apply to contractor employees. Any discipline issues concerning contractor employees should be referred immediately to the individual's manager. If criminal activity is suspected, contact the supporting legal staff for guidance.

(9) (JAG) Removal of contractor employees from the theater for disciplinary reasons, and their replacement, are at the employing contractor's expense.

(10) (JAG) Contractor employees will not be placed in a position that jeopardizes their status as a civilian authorized to accompany the force in the field, such as participation in offensive operations.

d. <u>Government-Furnished Support</u>.

(1) (G-4 and JAG) All GFE, including vehicles, are furnished to a contractor IAW the specific terms and conditions of the applicable contract.

(2) (G-4) Units or organizations providing GFE, including vehicles, will ensure that proper property accountability procedures contained in AR 710-2 and AR 735-5 are followed.

(3) (G-4) The requiring unit or activity provides or arranges for any maintenance and fuel support for GFE provided to a contractor.

(4) (JAG) Government-furnished services, such as telecommunications resources, maintenance and fuel support to contractor-furnished equipment and vehicles, is provided IAW the specific terms and conditions of the applicable contract.

(5) (Engineer) All facilities utilized by external support and system contractors is provided or arranged for by the government, through the appropriate staff engineer. Contractors specify facility requirements during contract negotiations. Theater support contractors utilize their existing facilities unless additional facilities are required. In that case, these requirements are identified during contract negotiations so the government can make arrangements for them.

(6) (G-4) Government-furnished transportation is provided to all external and system contractors, except the LOGCAP contractor, IAW the specific terms and conditions of the applicable contract and paragraph 3c (Deployment/Redeployment) of this annex. The LOGCAP contractor provides its own transportation support for this operation.

(7) (G-4) Lodging and subsistence support is furnished to all external and system contractor employees on a non-reimbursable basis. Theater support contractor employees receive non-reimbursable subsistence support when working away from their normal source of rations.

(8) (G-1) All US citizen external support and system contractor employees are authorized postal service through the MPS. Local nationals and non-US contractor employees receive postal support as arranged by the employing contractor.

(9) (Surgeon) Health care, including medical, dental, and pharmaceutical, is furnished to all external and system contractor employees on a non-reimbursable basis. Theater support contractor employees are NOT authorized government-furnished health care, except in job-related situations, and only on an emergency basis.

(10) (Surgeon) Occupational health and preventative medicine inspections, screenings, and remedial treatment is provided to all contractor employees and animals, regardless of their status.

(11) (Chaplain) Religious support is furnished to all contractor employees IAW the religious support annex of this OPORD.

(12) (JAG) Legal support to contractor employees is the responsibility of the employing contractor.

(13) (G-4 and G-1) The government provides mortuary affairs support to ALL contractor employees. Notification of next of kin for deceased contractor employees is the responsibility of the government for US citizen employees and the employing contractor for non-US citizen employees.

<u>e. Force Protection.</u>

(1) (G-3) Currently, there are no restrictions regarding where supporting contractors may provide support within the AO. The Third (US) Army G-3 will put out any changes to this policy.

(2) (G-3) Per local risk assessments, military forces provide necessary force protection for contractor employees working within their AO.

(3) (G-3) All contractors supporting this operation comply with force-protection guidelines issued by this headquarters and contained in this annex, including travel security requirements and off-limits restrictions. Commanders may waive the two-driver rule for contractor employees within their AO on a case-by-case basis.

(4) (G-3 and G-4) All contractors coordinate local travel of their employees within the AO with the appropriate operations and transportation staff.

(5) (G-3) Contractor employees are NOT authorized to carry weapons. Contractor employees violating this directive will be immediately removed from the theater and replaced at the employing contractor's expense.

(6) (G-3 and G-4) All contractor employees are issued and trained to properly wear OCIE and CDE.

(7) (G-3) Contractor employees supporting this operation are NOT authorized to wear BDUs/DCUs or any other military-looking uniform with the exception of individual protective gear.

(8) (G-3) All contractor employees receive training in the Geneva convention, code of conduct, status of forces agreements, safety, security, and NBC training.

(9) (Engineer) All external support and system contractor employees reside in government-furnished lodging. Theater support contractor employees reside at their normal residence unless required to work in a location too distant, in which case government-furnished lodging is used.

4. SERVICE SUPPORT: See Basic OPORD.

5. COMMAND AND SIGNAL

a. Command.

- The G-3, Third (US) Army, ensures that contractors are properly integrated into the operation. The PARC, Third (US) Army, is responsible for the overall management of theater support contractors. The senior support commands utilizing contractor support and associated Third (US) Army staff manage contractor support within their functional area.

- CORs are appointed for all contracts supporting this operation.

b. Signal.

- All contractors supporting this operation must interface with military communication systems. Contractors will comply with all security requirements and procedures.

- The supporting signal unit commander assists contractors in establishing interface with military communications systems.

ACKNOWLEDGE:

OFFICIAL:

BARNES
G4

LEWI
GEN

Appendix C

EXAMPLE LETTER OF AUTHORIZATION/IDENTIFICATION

DEPARTMENT OF THE ARMY
HEADQUARTERS, US ARMY COMMUNICATIONS-ELECTRONICS COMMAND
AND FORT MONMOUTH
FORT MONMOUTH, NEW JERSEY 07703-5000

AMSEL-LC-RE-FM **15 Nov 00**

TO: Whom It May Concern

SUBJECT: Letter of Authorization/Identification

Mr. John Doe, 555-55-5555, equivalent grade of GS-12, the bearer of this letter, is an employee of XYZ, Inc. which has a contract with this agency under government contract number GS-02-T98-CJA-0105. The contract period of performance is from 1 October 2000 to 30 September 2001. **The named bearer is not eligible or authorized to use available travel discounts in accordance with government contracts or agreements. Government contract city pair fares are not available to contractors.** Additionally, the named bearer is authorized to use government-furnished transportation if commercially available transportation is not available or is not cost effective.

In accordance with the Joint Travel Regulation, VOL II, Mr. Doe is entitled to all discounts (hotel, rental car, etc.) given to civilian personnel working for the Department of Defense.

Mr. Doe is additionally authorized to utilize government-furnished services and facilities, to include post exchange, commissary, care and treatment at medical and dental facilities, and use of government messing, billeting, and MWR facilities.

Mr. Doe will depart Fort Monmouth, NJ, for Fort Benning, GA; Seckenheim, Germany; Scopje, Macedonia; and Kosovo on or about 20 November 2000, and return to Fort Monmouth, NJ, after outprocessing through Fort Benning, GA, on or about 20 November 2001. He will support KFOR, TFF, and NATO missions as outlined by the 5th Signal Command. Ultimate unit of assignment is 17th Signal Battalion. Mr. Doe has a current security clearance of SECRET.

Excess baggage authorized: 3 bags, 60 pounds each.

Marvin T. Schultz
Contracting Officer

2112020 62-7510 P4200000 S28043
42 3808.00000.0000.RFZZZ.1

Appendix D

HEALTH ASSESSMENT QUESTIONNAIRES

This appendix contains two recommended health assessment questionnaires that may be used by medical personnel when conducting predeployment and redeployment medical screening. These questionnaires, filled out by the concerned contractor employee, are used along with provided medical records to assess whether or not an individual is medically fit to deploy to an AO and to assess possible long-term health impacts upon their return.

Pre-Deployment Health Assessment Questionnaire

INSTRUCTIONS: Please read each question carefully before marking your selections. Provide a response for each question. If you do not understand a question, ask the medical administrator conducting the medical screening.

DEMOGRAPHICS

Today's date (mm/dd/yy)

Anticipate deployment to

Last name

First name

Middle initial

Social security number
(or other identification number
if not a US citizen)

Date of birth

Gender
[ ] Male
[ ] Female

Supported military component (select only one)
[ ] Army
[ ] Air Force
[ ] Navy

[ ] Marine Corps

Citizenship category (select only one)
[ ] US
[ ] TCN (list nationality) ____________________
[} Local national (list nationality) ____________________

HEALTH ASSESSMENT

1. Would you say your health in general is
   [ ] Excellent
   [ ] Very good
   [ ] Good
   [ ] Fair*
   [ ] Poor*

2. During the past 90 days, how often did you seek medical care for an illness?
   [ ] Never
   [ ] Once
   [ ] 2-4 times*
   [ ] 5 or more times*

3. During the past 90 days, how often did you seek medical care for an injury?
   [ ] Never
   [ ] Once
   [ ] 2-4 times*
   [ ] 5 or more times*

4. During the past 90 days, how many days of work did you miss due to illness or injury?
   [ ] None
   [ ] 1-6 days
   [ ] 7-15 days*
   [ ] 16 or more days*

5. During the past year, did you stay in any hospital or medical facility overnight or longer?
   [ ] No
   [ ] Yes*

6. Are you currently on light duty or other work restrictions?
   [ ] No
   [ ] Yes*

7. Do you currently have any dental problems?
   [ ] No
   [ ] Yes*

8. Do you currently have any medical problems?
   [ ] No
   [ ] Yes*

9. Do you have any allergies?
   [ ] No
   [ ] Yes*

10. Are you regularly taking any medications? (select all that apply)
   [ ] No, I am not taking any medications
   [ ] Over-the-counter medications
   [ ] Prescription medication*
   [ ] Birth control pills
   [ ] Vitamins
   [ ] Herbal supplements
   [ ] Malaria pills*

11. If you are taking prescription medications or birth control pills, do you have enough to last 90 days?
   [ ] No*
   [ ] Yes
   [ ] Not applicable

12. (For females) What was the result of your last PAP smear? Date of last WWE/PAP_________
   [ ] Normal
   [ ] Abnormal*
   [ ] Don't know*

13. (For females) Are you pregnant?
   [ ] No
   [ ] Yes
   [ ] I am not sure*

QUESTIONS 14-16 PERTAIN TO YOUR MENTAL HEALTH, WHICH INCLUDES ALCOHOL PROBLEMS, STRESS, DEPRESSION, AND EMOTIONAL PROBLEMS.

14. During the last 30 days, how many days was your mental health not good?
   [ ] None
   [ ] 1-5 days
   [ ] 6-10 days
   [ ] 11-15 days*
   [ ] 16 or more days*

15. During the last 30 days, how many days did your mental health keep you from your usual activities, such as self-care, work, or recreation?
   [ ] None
   [ ] 1-5 days
   [ ] 6-10 days
   [ ] 11-15 days*
   [ ] 16 or more days*

16. During the past year, have you sought counseling or care for your mental health?
   [ ] No
   [ ] Yes*

17. During the past 30 days, have you seriously considered injuring yourself or others?

[ ] No
[ ] Yes*

18. Have you ever suffered or sought treatment for any heat related injury such as heat stroke?
[ ] No
[ ] Yes*

19. Have you ever suffered or sought treatment for a cold injury such as frost bite or immersion foot?
[ ] No
[ ] Yes*

20. Do you currently have any questions or concerns about your health?
[ ] No
[ ] Yes*

21. Do you have concerns about exposure (such as environmental or work-related) that may affect your health?
[ ] No
[ ] Yes*

*Denotes that health care provider must follow-up!

END OF QUESTIONNAIRE

Pre-Deployment Health Provider Review
(for health provider only)

REVIEW

Indicate status of each of the following:

Yes	No	N/A	
[ ]	[ ]	[ ]	Medical threat briefing completed
[ ]	[ ]	[ ]	Medical information sheet distributed
[ ]	[ ]	[ ]	Pre-deployment serum specimen collected
[ ]	[ ]	[ ]	Exposure concerns reviewed (if yes, indicate type of exposure(s) reviewed)

X	Exposure Type
[ ]	Environment (air/soil/water)
[ ]	NBC warfare risks
[ ]	Immunizations
[ ]	Chemoprophalaxis
[ ]	Infectious diseases
[ ]	Occupational exposures (chemical, physical, biological)
[ ]	Other (list)______________________________

Yes	No	N/A	
[ ]	[ ]	[ ]	Referred for further evaluation(s) (if yes, indicate type(s) of referral and disposition(s))

X	Referral Type
[ ]	Physical examination
[ ]	Dental examination

[ ] Infectious and parasitic diseases
[ ] Neoplasm
[ ] Endocrine-nutrition and metabolic disorders and immunity disorders
[ ] Diseases of the blood and blood-forming organs
[ ] Mental disorders
[ ] Diseases of the nervous system and sense organs
[ ] Diseases of the circulatory system
[ ] Diseases of the respiratory system
[ ] Diseases of the digestive system
[ ] Diseases of the genitourinary system
[ ] Diseases or conditions of the reproductive system
[ ] Diseases of the skin and subcutaneous tissue
[ ] Diseases of the musculoskeletal systemand connective tissue
[ ] Symptoms and signs of ill-defined conditions
[ ] Injury and poisoning
[ ] Other, list____

Final medical disposition: [ ] Deploy [ ] Not Deploy (If not deployable, explain why)

I certify that this screening process has been completed.

Medical provider's signature and stamp:

Date:

Redeployment Health Assessment Questionnaire

INSTRUCTIONS: Please read each question carefully before marking your selections. Answer each question. If you do not understand a question, ask the medical administrator conducting the medical screening.

DEMOGRAPHICS

Today's Date (mm/dd/yy)

Anticipate deployment to

Last name

First name

Middle initial

Social security number
(or other identification number
if not a US citizen)

Date of birth

Gender
[ ] Male
[ ] Female

Supported military component (select only one)
[ ] Army
[ ] Air Force
[ ] Navy
[ ] Marine Corps

Citizenship category (select only one)
[ ] US
[ ] TCN (list nationality) ________________________
[} Local national (list nationality) _________________________

HEALTH ASSESSMENT

1. Would you say your health in general is—
[ ] Excellent
[ ] Very good
[ ] Good
[ ] Fair*
[ ] Poor*

2. Compared to before you were deployed, would you say your health in general is—
   [ ] Much better now
   [ ] Somewhat better now
   [ ] About the same now
   [ ] Somewhat worse now*
   [ ] Much worse now*

3. During this deployment, how often did you seek medical care for an illness?
   [ ] Never
   [ ] Once
   [ ] 2-4 times
   [ ] 5 or more times*

4. During this deployment, how often did you seek medical care for an injury?
   [ ] Never
   [ ] Once
   [ ] 2-4 times
   [ ] 5 or more times*

5. During this deployment, how many days of work did you miss due to illness?
   [ ] None
   [ ] 1-6 days
   [ ] 7-15 days*
   [ ] 16 days or more*

6. During this deployment, how many days of work did you miss due to injury?
   [ ] None
   [ ] 1-6 days
   [ ] 7-15 days*
   [ ] 16 days or more*

7. During this deployment, did you stay in any hospital or medical facility overnight or longer?
   [ ] No
   [ ] Yes*

8. Are you currently on light duty or other work restrictions??
   [ ] No
   [ ] Yes*

9. Do you currently have any dental problems?
   [ ] No
   [ ] Yes*

10. Do you currently have any medical problems?
   [ ] No
   [ ] Yes*

11. Are you regularly taking any medications? (select all that apply)
   [ ] No, I am not taking any medications
   [ ] Over-the-counter medications
   [ ] Prescription medication*
   [ ] Birth control pills

[ ] Malaria pills*

12. (For females) Are you pregnant?
    [ ] No
    [ ] Yes
    [ ] I am not sure*

QUESTIONS 13-15 PERTAIN TO YOUR MENTAL HEALTH, WHICH INCLUDES ALCOHOL PROBLEMS, STRESS, DEPRESSION AND EMOTIONAL PROBLEMS

13. During the last 30 days, how many days was your mental health not good?
    [ ] None
    [ ] 1-5 days
    [ ] 6-10 days
    [ ] 11-15 days*
    [ ] 16 or more days*

14. During the last 30 days, how many days did your mental health keep you from your usual activities, such as self-care, work, or recreation?
    [ ] None
    [ ] 1-5 days
    [ ] 6-10 days
    [ ] 11-15 days*
    [ ] 16 or more days*

15. During this deployment, have you sought counseling or care for your mental health?
    [ ] No
    [ ] Yes*

16. During this deployment, have you seriously considered injuring yourself or others?
    [ ] No
    [ ] Yes*

17. Do you have concerns about exposure (such as environmental or work-related) during this deployment that you feel may affect your health?
    [ ] No
    [ ] Yes*

18. During this deployment, have you suffered or sought treatment for any heat-related injury such as heat exhaustion?
    [ ] No
    [ ] Yes*

19. During this deployment, have you suffered or sought treatment for a cold injury such as frost bite or immersion foot?
    [ ] No
    [ ] Yes*

20. Do you currently have any questions or concerns about your health?
    [ ] No
    [ ] Yes*

* Denotes that health care provider must follow-up!

END OF QUESTIONNAIRE

Redeployment Health Provider Review
(for health provider only)

REVIEW

Indicate status of each of the following:

Yes	No	N/A	
[ ]	[ ]	[ ]	Medical threat briefing completed
[ ]	[ ]	[ ]	Medical information sheet distributed
[ ]	[ ]	[ ]	Pre-deployment serum specimen collected
[ ]	[ ]	[ ]	Exposure concerns reviewed (if yes, indicate type of exposure(s) reviewed)

X	Exposure Type
[ ]	Environment (air/soil/water)
[ ]	NBC warfare risks
[ ]	Immunizations
[ ]	Chemoprophalaxis
[ ]	Infectious diseases
[ ]	Occupational exposures (chemical, physical, biological)
[ ]	Others, list________________________________

Yes	No	N/A	
[ ]	[ ]	[ ]	Referred for further evaluation(s) (if yes, indicate type(s) of referral and disposition(s))

X	Referral Type
[ ]	Physical examination
[ ]	Dental examination
[ ]	Infectious and parasitic diseases
[ ]	Neoplasm
[ ]	Endocrine-nutrition and metabolic disorders and immunity disorders
[ ]	Diseases of the blood and blood-forming organs
[ ]	Mental disorders
[ ]	Diseases of the nervous system and sense organs
[ ]	Diseases of the circulatory system
[ ]	Diseases of the respiratory system
[ ]	Diseases of the digestive system
[ ]	Diseases of the genitourinary system
[ ]	Diseases or conditions of the reproductive system
[ ]	Diseases of the skin and subcutaneous tissue
[ ]	Diseases of the musculoskeletal system and connective tissue
[ ]	Symptoms and signs of ill-defined conditions
[ ]	Injury and poisoning
[ ]	Other, list____________________________________

Final medical disposition: [ ] Deploy [ ] Not Deploy (If not deployable, explain why)

I certify this screening process has been completed.

Medical provider's signature and stamp:

Date:

Glossary

AAFES	Army/Air Force Exchange Service
ABCS	Army Battle Command Systems
ACO	Administrative Contracting Officer
AIT	automated information technology
AMC	Air Mobility Command
AO	area of operation
AOR	area of responsibility
APOD	aerial port of debarkation
APOE	aerial port of embarkation
AR	Army Regulation
ARB	Acquisition Review Board
ARINC	Aeronautical Radio Inc.
ARFOR	Army Forces
ASA(ALT)	Assistant Secretary of the Army, (Acquisition, Logistics and Technology)
ASCC	Army Service Component Commander
AT	antiterrorist
BCF	Basic Core Formulary
C2	command and control
CDE	chemical defense equipment
CLPSB	combatant commanders logistics procurement support board
CLS	contracted logistic support

COMSEC	communications security
CONUS	Continental United States
COA	Courses of Action
COR	contracting officer's representative
COSCOM	corps support command
CRAF	Civil Reserve Air Fleet
CREST	Corps Real Estate Support Team
CRC	CONUS Replacement Center
CS	combat support
CSAR	Combat search and rescue
CSS	combat service support
DAC	Department of the Army civilian
DACG	departure airfield control group
DCMA	Defense Contract Management Agency
DCSLOG	Deputy Chief of Staff, Logistics
DFARS	Defense Federal Acquisition Supplement
DHP	Defense Health Program
DISCOM	division support command
DLA	Defense Logistics Agency
DOB	date of birth
DOD	Department of Defense
DODI	Department of Defense Instruction
EAC	echelons above Corps
EAD	echelons above division

Emergency Care	Medical treatment of patients with severe life-threatening or potentially disabling conditions resulting from accident or illness of sudden onset. These conditions necessitate immediate care to prevent undue suffering or loss of life or limb. Dental treatment for relief of painful or acute condition.
FM	Field Manual
FP	force protection
FORSCOM	Forces Command
FTX	field training exercise
GFE	government furnished equipment
GPMRC	Global Patient Movement Requirements Center
HCA	Head of Contracting Activity: A general officer, usually the senior commander in the theater, who provides overall contracting guidance; serves as the approving authority for contracting; and appoints the PARC. All contracting authority in a theater flows from the HCA to the PARC.
HQDA	Headquarters, Department of the Army
Host Nation	A nation which receives the forces and/or supplies of allied nations and/or organizations to be located on, to operate in, or to transit through its territory.
HNS	host nation support: civil and/or military assistance rendered by a nation to foreign forces within its territory during peacetime, times of crisis, emergency, or war, based upon agreements mutually

	concluded between nations.
ICS	interim contracted logistic support
IDS	individual deployment site
IPB	Intelligence preparation of the battlefield
IPE	individual protective equipment
ITV	in-transit visibility
JAG	Judge Advocate General
JARB	Joint Acquisition Review Board
JCS	Joint-Chiefs-of-Staff
JCSP	joint contracting support plan
JFC	joint forces command
Joint Operations	Operations carried on by two or more of the Armed Forced of the United States (Army, Navy, Air Force).
JOPES	Joint Operation Planning and Execution System
JP	Joint Publication
JTF	Joint Task Force
LAO	logistics assistance officer
LOC	lines of communication
LOGCAP	Logistics Civil Augmentation Program
LPB	Logistics preparation of the battlefield
LSE	logistics support element
MACOM	major army command
MCX	Marine Corps Exchange
MDMP	Military decision-making process
METT-TC	mission, enemy, terrain, troops, time available, and civil considerations

MMR	measles, mumps, rubella
MOG	maximum on ground
MPS	Military Postal Service
MSE	mobile subscriber equipment
MTF	medical treatment facility
MTMC	Military Traffic Management Command
MTP	Mission Training Plan
MTW	Major theater war
MWR	morale, welfare and recreation
Nation Assistance	Civil and/or military assistance rendered to a nation by foreign forces within that nation's territory during peacetime, crises or emergencies, or war, based on agreements mutually concluded between nations.
NBC	nuclear, biological, and chemical
NCA	National Command Authorities
NEX	Navy Exchange
NRP	non-unit related personnel
OCIE	organizational clothing and individual equipment
OPCON	operational control
OPLAN	operations plan
OPORD	operation order
OPSEC	operations security

PARC	Principal Assistant Responsible for Contracting: a special staff officer, is the ASCC or mission commander's senior Army acquisition advisor responsible for planning and managing all Army contracting functions within the theater.
PCO	Procuring Contracting Officer
PEO	Program Executive Officer
PM	Program Manager
POC	point-of-contact
POD	port of debarkation
POE	port of embarkation
Routine care	Routine medical care is nonemergency care that is required and medically indicated. Routine dental is all care necessary to maintain dental health and function other than care of an emergency or elective nature.
RSO&I	reception, staging, onward movement, and integration
SERE	Survival, escape, resistance and evasion
SITA	Societe International de Telecommunications Aeronautiques
SJA	Staff Judge Advocate
SOO	statement of objectives
SOW	statement of work
SPOD	seaport of debarkation
STX	situational training exercise

TACC	Tanker Airlift Command Center
TALCE	Tanker Airlift Control Element
TCO	Terminating Contracting Officer
TPFDD	time-phased force and deployment data
TPMRC	Theater Patient Movement Requirements Center
TPS	tactical personnel system
TSC	theater support command
TSP	training support package
UCMJ	Uniform Code of Military Justice
UIC	unit identification code
ULN	unit line number
UMT	Unit Ministry Team
US	United States
UTC	unit type code
USACE	US Army Corps of Engineers
USAMC	US Army Materiel Command
USAREUR	US Army Europe
USARPAC	US Army Pacific
USO	United Services Organization
USTRANSCOM	US Transportation Command

Bibliography

AMC Pam 700-30, *Logistics Civil Augmentation Program*, December 1997.

AMC Pam 715-18, *USAMC Contractor Deployment Guide for Contracting Officers*, July 1996.

AR 40-3, *Medical, Dental, and Veterinary Care*, July 1999.

AR 40-562, *Immunizations and Chemoprophylaxis*, November 1995.

AR 600-8-1, *Army Casualty Operations/Assistance/Insurance*, October 1994.

AR 600-8-14, *Identification Cards for Members of the Uniformed Services, Thei Family Members, and Other Eligible Personnel*, March 1998.

AR 700-4, *Logistics Assistance*, June 1995.

AR 700-127, *Integrated Logistics Support*, November 1995.

AR 710-2, *Inventory Management Supply Police Below the Wholesale Level*, October 1997.

AR 700-137, *Logistics Civil Augmentation Program (LOGCAP)*, December 1985.

AR 715-9, *Contractors Accompanying the Force*, October 1999.

AR 735-5, *Policies and Procedures for Property Accountability*, January 1998.

Army Federal Acquisition Regulation Supplement Manual No. 2, *Contingency Contracting*, December 1993.

CJCSM 3122.02 *Crisis Action Time-Phased Force and Development Data Development and Deployment Execution*, Volume III, Enclosure G, Appendix A 25 May 2001.

DA Pam 27-1, *Treaties Governing Land Warfare*, December 1956.

DA Pam 27-1-1, *Protocols to the Geneva Conventions of 12 August 1949*, September 1979.

DA Pam 690-47, *DA Civilian Employee Deployment Guide*, November 1995.

DA Pam 690-80, *Use and Administration of Local Civilians in Foreign Areas During Hostilities*, February 1971.

DA Pam 715-16, *Contractor Deployment Guide*, April 1998.

DA Policy Memorandum, *Contractors on the Battlefield*, December 1997.

DOD Acquisition Deskbook Supplement, *Contractor Support in the Theater of Operations*, March 2001.

DOD Directive 2310.2, *Personnel Recovery*, December 2000.

DOD Directive 5000.1, *Defense Acquisition*, March 1996.

DOD Instruction 1000.1, *Identification Cards Required by the Geneva Conventions*, January 1974.

DOD Instruction 1000.13, *Identification Cards for Members of the Uniformed Services, Their Dependents, and Other Eligible Individuals*, December 1997.

DOD Instruction 2310.4, *Repatriation of Prisoners of War (POW), Hostages, Peacetime Government Detainees and Other Missing or Isolated Personnel*, November 2000.

DOD Instruction 3020.37, *Continuation of Essential DoD Contractor Services During Crises*, November 1990.

DOD O-2000.12-H, *DOD Antiterrorism Handbook*, 2002. (draft)

HQDA Letter 1-01-1, *Force Health Protection (FHP): Occupational and Environmental Health (OEH) Threats,* June 2001

FM 3-0, *Operations,* June 2001.

FM 4-0 (100-10), *Combat Service Support,* October 2001. (final draft)

FM 5-100, *Engineer Operations*, February 1996.

FM 8-10, Health Support In A Theater of Operations, March 1991.

FM 10-1, Quartermaster Principles, August 1995.

FM 12-6, *Personnel Doctrine*, September 1994.

FM 14-100, Financial Management Operations, May 1997.

FM 27-10, The Law of Land Warfare, July 1956.

FM 27-100, *Legal Operations*, September 1991.

FM 54-8, Division Materiel Management Center, January 1984.

FM 54-23, Materiel Management Center-Corps Support Command, December 1984.

FM 54-30, Corps Support Groups, June 1993.

FM 55-10, Movement Control in a Theater of Operations, December 1992.

FM 55-65, *Strategic Deployment*, May 1989.

FM 63-2, Division Support Command, May 1991.

FM 63-3, *Corps Support Command,* September 1993.

FM 63-11, *Logistics Support Element*, October 1996.

FM 71-100, *Division Operations*, August 1996.

FM 100-7, Decisive Force, The Army in Theater Operations, May 1995.

FM 100-8, The Army in Multinational Operations, November 1997.

FM 100-10-2 (4-100.2), Contracting Support on the Battlefield, March 2000.

FM 100-14, *Risk Management*, April 1995.

FM 100-15, *Corps Operations*, October 1996.

FM 100-16, Army Operational Support, May 1995.

FM 100-17, Mobilization, Deployment, Redeployment, and Demobilization, April 1994.

FM 100-17-1, Army Pre-Positioned Afloat Operations, July 1996.

FM 100-17-2, *Army Pre-Positioned Land*, February 1999.

FM 100-17-3, Reception, Staging, On-Ward Movement, and Integration (RSOI), March 1999.

FM 100-17-4, *Deployment: Fort to Port*, April 2001. (draft)

FM 100-17-5, *Redeployment*, September 1999.

FM 100-23, *Peace Operations*, December 1994.

FM 100-25, *Army SOF Operations*, December 1991.

FM 101-5, Staff Organization and Operations, May 1997.

Geneva Convention for the Amelioration of the Condition of the Wounded and Sick in Armed Forces in the Field, August 1949.

Geneva Convention Relative to the Treatment of Prisoners of War, August 1949.

Joint Pub 0-2, Unified Action Armed Forces (UNAAF), July 2001.

Joint Pub 1-0, Joint Doctrine for Personnel Support to Joint Operations, November 1998.

Joint Pub 1-02, DOD Dictionary of Military and Associated Terms, April 2001.

Joint Pub 3-0, *Doctrine for Joint Operations*, September 2001.

Joint Pub 3-07, Joint Doctrine for Military Operations Other than War, June 1995.

Joint Pub 3-07.3, *JTTP for Peace Operations*, February 1999.

Joint Pub 3-07.6, JTTP for Foreign Humanitarian Assistance, August 2001.

Joint Pub 3-10, Joint Doctrine for Rear Operations, May 1996.

Joint Pub 3-10, *JTTP for Base Defense*, July 1996.

Joint Pub 4-0, Doctrine Support for Logistics Support for Joint Operations, April 2000.

Joint Pub 4-01.4, JTTP for Joint Theater Distribution, August 2000.

Joint Pub 4-01.8, JTTP for Joint Reception, Staging, On-ward Movement, and Integration, June 2000.

Joint Pub 4-05, *Mobilization*, June 1995.

Joint Pub 4-07, Joint Tactics, Techniques, and Procedures for Common User Logistics During Joint Operations, June 2001.

Made in United States
Troutdale, OR
11/03/2026

56844380R00069